T0331727

CONCEPTS IN
RELATIVISTIC
DYNAMICS

CONCEPTS IN
RELATIVISTIC
DYNAMICS

Lawrence Horwitz

Tel Aviv University, Israel

World Scientific

NEW JERSEY · LONDON · SINGAPORE · BEIJING · SHANGHAI · HONG KONG · TAIPEI · CHENNAI · TOKYO

Published by

World Scientific Publishing Co. Pte. Ltd.

5 Toh Tuck Link, Singapore 596224

USA office: 27 Warren Street, Suite 401-402, Hackensack, NJ 07601

UK office: 57 Shelton Street, Covent Garden, London WC2H 9HE

Library of Congress Control Number: 2023932605

British Library Cataloguing-in-Publication Data
A catalogue record for this book is available from the British Library.

CONCEPTS IN RELATIVISTIC DYNAMICS

ISBN 978-981-120-731-0 (hardcover)
ISBN 978-981-120-732-7 (ebook for institutions)
ISBN 978-981-120-733-4 (ebook for individuals)

For any available supplementary material, please visit
https://www.worldscientific.com/worldscibooks/10.1142/11474#t=suppl

Desk Editor: Carmen Teo Bin Jie

Typeset by Stallion Press
Email: enquiries@stallionpress.com

Printed in Singapore

Dedicated to my son, Benjamin Horwitz, Professor of Biology,
Technion, Haifa, Israel, and my daughters, Dorothy Simon,
teacher in Edison, N. J., and Debbie Wexler,
teacher in Kedumim, Israel.

Acknowledgments

I would like to thank Constantin Piron for his collaboration in developing the main ideas discussed in this book during visits to the University of Geneva, hosted initially by J.M. Jauch, and for many of the discussions and collaborations which followed. I am grateful to Stephen L. Adler for his warm hospitality at the Institute for Advanced Study in Princeton over a period of several visits during which, among other things, much of the research on this subject was done. I am also grateful to William C. Schieve ($Z''L$) for his collaborations over many years, particularly in the development of relativistic statistical mechanics, and for the hospitality he and his wife, Florence, offered me at the University of Texas at Austin during many visits. I would also like to thank Fritz Rohrlich for his hospitality at Syracuse University and for many discussions on relativistic quantum theory and for his collaboration in our study of the constraint formalism.

I am also particularly indebted to Rafael I. Arshansky for his collaboration on developing many of the basic ideas, such as the solution of the two body bound state problem, the Landau-Peierls relation and many aspects of scattering theory, as well as Yeshiyahu Lavie ($Z''L$), who fell in the first Lebanon-Israel war; he played an essential role in the development of many of the basic ideas in scattering theory as well. I am grateful to Y. Rabin who worked with me on the first analyses of the phenomenon of interference in time in 1976, to Ori Oron for his contributions to the application of the eikonal approximations to the relativistic equations and to the relativistic generalization of Nelson's approach to the quantum theory through the use of Brownian motion, and Nadav Shnerb for the development of the second quantized formalism with the help and guidance of Kurt Haller (Z″L), to Igal Aharonovich for his intensive work on the classical electromagnetic properties of the relativistically charged particle and the associated five dimensional fields, to Martin Land for his many invaluable contributions to the theory and many of its applications, to Avi Gershon for his invaluable contribution to the application of the theory to a geometrical approach to the dark matter problem, to Andrew Bennett, who found the way to compute the anomalous magnetic moment of the electron in this framework, as well as many discussions, to Meir Zeilig-Hess for his contribution to the tensor representations for spin, and to many other students and colleagues that have contributed to the effort to understand the structure and consequences of the theory discussed here.

I am especially grateful to Professor Asher Yahalom and Ariel University for organizing and hosting an international conference "General Relativity, Quantum Mechanics, and Everything in Between, Celebrating 92 Springs of Professor Lawrence P. Horwitz", which took place on April 25 and 26, 2022, at Ariel University.

I am very grateful to Carmen Teo Bin Jie, of World Scientific, for her careful editing and assistance in preparing the manuscript for publication.

Finally, I wish to thank my wife, Ruth, for her patience and support in the many years during which the research was done in developing and exploring the consequences of the approach to relativistic mechanics and quantum mechanics that is discussed in this book.

Contents

Chapter 1

Introduction

In the past fifty years there have been significant developments in the theory of relativistic mechanics, both classical and and quantum. It is the purpose of this book to explain and emphasize the conceptual foundations of these developments with minimal use of advanced mathematical techniques, so that these ideas will be accessible to a wide range of readers, students and researchers.

Before the beginning of the twentieth century it was widely believed that the observed electromagnetic waves must travel, as sound waves, in some medium. Maxwell, who formulated the theory of electromagentism, believed that the waves predicted by the theory are carried by a medium he called the "aether", referred to later simply as "ether". Michelson and Morley then tried to measure the speed of the Earth through this medium by constructing an interferometer that detects interference between light waves travelling in two perpendicuar directions superimposed by mirrors, an arrangement very sensitive to the speed of light. If the Earth is travelling through the ether in the direction of the incoming light, its wavelength would appear to be shorter (arriving at greater speed), and when the Earth is going in the opposite direction, the wavelength would be longer, and the speed of light smaller. They found, to high accuracy, that there was no difference. Furthermore, the speed of the two arms of the apparatus through the ether should be different, but rotating by ninety degrees had no such effect.

To explain this null result, Lorentz supposed that there was a deformation of the material of the interferometer, and wrote a formula for the deformation that involved both the time interval between the waves and the displacement of the material. The compensation, which could account for the null result, called the "Lorentz contraction", was written as

$$\Delta x' = \frac{\Delta x - v\Delta t}{\sqrt{1 - \frac{v^2}{c^2}}}, \tag{1.1}$$

where c is the velocity of light in the ether. Einstein perceived that this formula expresses simply that the velocity of light, measured in any inertial frame, would have the same value; there is no need for an ether, or a compression of material in the measuring apparatus.

If the velocity of light is the same in any state of steady motion, then the distance ℓ that the light travels in a time t must be

$$\Delta \ell = c\Delta t, \tag{1.2}$$

and, for the same physical process measured with apparatus in a different state of motion, allowing for a change in the time interval as well, the corresponding result should be

$$\Delta \ell' = c\Delta t', \tag{1.3}$$

for the same value of c.

To establish a relation between $\Delta\ell$, Δt and $\Delta\ell'$, $\Delta t'$ constituting a *transformation* between the two descriptions, Einstein imposed the requirement that

$$c^2\Delta t^2 - \Delta x^2 = c^2\Delta t'^2 - \Delta x'^2. \tag{1.4}$$

It actually follows from (1.2) and (1.3) that both sides vanish, but we wish to construct a non-trivial transformation of coordinates that is sufficient to preserve the measured light speed with any apparatus in steady motion. If both sides vanish identically, a simple scale transformation would suffice, and we would not recover the formula of Lorentz. Furthermore, the relations (1.2) and (1.3) are for light. The transformation of coordinates defined by (1.4), is a result that should apply to our perception of space and time for massive particles as well; as we shall see later, in this case $c^2\Delta t^2 - \Delta x^2$, in general, does not vanish, but the transformation law derived from (1.4) remains valid.

In the following, we derive Lorentz's transformation directly from the statement that the velocity of light is measured to be the same as measured by any apparatus moving at any relative constant speed (inertial). We shall study the detection of signals received by such an apparatus in steady motion relative to a signal generating apparatus.

The transformation (1.1) discovered by Lorentz that successfully accounted for the null result of Michelson and Morley shows that both the time and space intervals must be affected by a linear transformation, so we start by writing

$$\Delta x^{0'} = A_0 \Delta x^0 + B_0 \Delta x^1$$
$$\Delta x^{1'} = A_1 \Delta x^0 + B_1 \Delta x^1,$$
(1.5)

where we have written $\Delta x^0 = c\Delta t$, for the interval between two signals, with c the velocity of light in the "ether" as Lorentz defined it, and Δx^1 for the space interval between the two emissions in the emitting system. It is sufficient for our present purpose to consider just one space dimension.

We then impose the relation (1.4) to obtain

$$-A_0{}^2 + A_1{}^2 = -1$$
$$-B_0{}^2 + B_1{}^2 = +1$$
(1.6)

and, for the vanishing of the cross-terms,

$$-A_0 B_0 + A_1 B_1 = 0.$$
(1.7)

Since the continuous act of moving to a relatively moving measuring apparatus should not change the direction of the flow of time (in particular for two signals emitted at the same point $\Delta x^1 = 0$), we must have, by (1.5) and (1.6),

$$A_0 = +\sqrt{1 + A_1{}^2} \geq 1.$$
(1.8)

Now consider a situation for which the two signals that we measure are emitted from the same point in the apparatus generating the signals so that $\Delta x^1 = 0$. Then, from (1.5),

$$\Delta x^{0'} = \gamma \Delta x^0, \tag{1.9}$$

where we have called, as is conventionally done,

$$A_0 \equiv \gamma. \tag{1.10}$$

For the second of (1.5) we have, in this case,

$$\Delta x^{1'} = A_1 \Delta x^0. \tag{1.11}$$

We then have that

$$A_1 = \frac{\Delta x^{1'}}{\Delta x^0} = \frac{\Delta x^{1'}}{\Delta x^{0'}} \frac{\Delta x^{0'}}{\Delta x^0} = -v\gamma, \tag{1.12}$$

where we have identified $-v \equiv \frac{\Delta x^{1'}}{\Delta x^{0'}}$ with the velocity of the signal emitter (as observed in the laboratory moving with $+v$) as detected in the moving apparatus (divided by c), and the second factor is defined by (1.9).

Now, with these results, by the first of (1.6) we have

$$-\gamma^2 + \gamma^2 v^2 = -1,$$

or

$$\gamma^2(v^2 - 1) = -1,$$

so that (determined to be positive above)

$$\gamma = +\frac{1}{\sqrt{1 - v^2}}. \tag{1.13}$$

The result (1.9) then corresponds to the well-known "time dilation".

By (1.5) we now have

$$\Delta x^{0'} = \gamma \Delta x^0 + B_0 \Delta x^1$$
$$\Delta x^{1'} = \gamma v \Delta x^0 + B_1 \Delta x^1. \tag{1.14}$$

Now, consider the case in which two signals are sent at equal time in the emitting system. Then, for $\Delta x^0 = 0$, equation (1.14) becomes

$$\Delta x^{0'} = \gamma B_0 \Delta x^1$$
$$\Delta x^{1'} = B_1 \Delta x^1. \tag{1.15}$$

For $D = 1$, as we argue below, $B_1 = \sqrt{1 - v^2}^{-1}$, and with this, the measurement of $\Delta x^{1'}$ implies that the interval Δx^1 has diminished (Lorentz contraction).

We now impose the condition that the transformation be *invertible*, so that an additional transformation in the reverse direction brings us back to the properties of the original signals emitted. One must therefore be able to solve (1.14) for Δx^0 and Δx^1 in terms of a linear combination of $\Delta x^{0'}$ and $\Delta x^{1'}$. The condition for finding the solution for the inverse is that the determinant of the coefficients be nonvanishing, *i.e.*,

$$A_0 B_1 - B_0 A_1 = D \neq 0, \tag{1.16}$$

or

$$\gamma B_1 - B_0 v \gamma = D. \tag{1.17}$$

Therefore,

$$B_1 = B_0 v + D/\gamma. \tag{1.18}$$

By (1.6),

$$B_1{}^2 = 1 + B_0{}^2,$$

so that

$$(B_0 + D/\gamma)^2 = 1 + B_0{}^2.$$

The solution to this quadratic equation is (using the definition of γ to simplify the result)

$$B_0 = -v\gamma D \pm \gamma\sqrt{D^2 - 1}, \tag{1.19}$$

and, with (1.18), we have B_0 and B_1 in terms of v and D.

The linear transformation for $\Delta x^0, \Delta x^1$ to $\Delta x^{0'}, \Delta x^{1'}$ is then

$$\Delta x^{0'} = \gamma\Delta x^0 + (-v\gamma D \pm \sqrt{D^2 - 1})\Delta x^1$$
$$\Delta x^{1'} = -v\gamma\Delta x^0 + (D\gamma \pm v\gamma\sqrt{D^2 - 1})\Delta x^1. \tag{1.20}$$

An additional transformation, corresponding to an additional change v' in relative velocity, must result in a linear transformation to the new total velocity v'' of the same form:

$$\Delta x^{0''} = \gamma''\Delta x^0 + (-v''\gamma''D'' \pm \sqrt{D''^2 - 1})\Delta x^1$$
$$\Delta x^{1'} = -v''\gamma''\Delta x^0 + (D''\gamma'' \pm v''\gamma'\sqrt{D''^2 - 1})\Delta x^1. \tag{1.21}$$

We now show that this requirement necessarily leads to $D = 1$. To construct the result of an additional transformation $x' \to x''$, we follow the same procedure as used to obtain (1.20), *i.e.*,

$$\Delta x^{0''} = \gamma'\Delta x^{0'} + (-v'\gamma'D' \pm \sqrt{D'^2 - 1})\Delta x^{1'}$$
$$\Delta x^{1''} = -v'\gamma'\Delta x^{0'} + (D'\gamma' \pm v'\gamma'\sqrt{D'^2 - 1})\Delta x^{1'}. \tag{1.22}$$

Now, substitute the results obtained for $\Delta x^{0'}, \Delta x^{1'}$ in (1.20) into (1.22) to give the overall transformation $x \to x''$ to obtain, after rearranging,

$$\Delta x^{0''} = \Delta x^0[\gamma'\gamma + (-v'\gamma'D' \pm \sqrt{D'^2 - 1})v\gamma]$$
$$+ \Delta x^1[\gamma'(-v\gamma D \pm \sqrt{D^2 - 1})$$
$$+ (v'\gamma'D' \pm \sqrt{D'^2 - 1})(D\gamma \pm v\gamma\sqrt{D^2 - 1})], \tag{1.23}$$

and

$$\Delta x^{1''} = \Delta x^0 [-v'\gamma'\gamma + (D'\gamma' \pm v'\gamma'\sqrt{D'^2 - 1})v\gamma]$$
$$+ \Delta x^1 [-v'\gamma'(-v\gamma D \pm \sqrt{D^2 - 1})$$
$$+ (D'\gamma' \pm v'\gamma'\sqrt{D'^2 - 1})(D\gamma \pm v\gamma\sqrt{D^2 - 1})]. \quad (1.24)$$

We can compare this result with (1.21) most easily for when the second transformation is small, *i.e.*, for small v'. We first remark that in this case, $A_0' = \gamma' \approx 1, A_1' == \gamma'v' \approx 0$, so that $D' \approx B_1'$. But, by (1.5), $B'_1 \approx 1$ in this case. The coefficient of Δx^0 in $\Delta x^{0''}$ is then approximately $\gamma \pm \sqrt{D'^2 - 1}v\gamma$. Although we have taken a small additional velocity v', for v large and close to unity, γ may be as large as we wish. The square root term in the coefficient of Δx^0 must then vanish, *i.e.* $D' = 1$ (positive by (1.5)).

Although we have reached the conclusion that the determinant of coefficients must be exactly unity for small additional motions, this algebraic property must be true for the initial transformation; furthermore, any finite shift in velocity can be constructed with a series of small steps, so our result is generally valid.

The transformation law now reads

$$\Delta x^{0'} = \gamma\Delta x^0 - v\gamma\Delta x^1$$
$$\Delta x^{1'} = -v\gamma\Delta x^0 + \gamma\Delta x^1, \quad (1.25)$$

or, replacing v by v/c and x^0 by ct, we now have the familiar form of Lorentz's transformation

$$\Delta t' = \gamma\Delta t - \frac{v}{c}\gamma\Delta x^1$$
$$\Delta x' = -v\gamma\Delta t + \gamma\Delta x, \quad (1.26)$$

for $\gamma = \frac{1}{\sqrt{1-(\frac{v}{c})^2}}$.

In this derivation of the Lorentz transformation we have not used any advanced mathematical techniques. The requirement that we proved by consistency arguments, that $D = 1$, also follows by asserting that compound transformations form a *group*; a mathematical structure with a set of elements (containing a unity element)

$a, b, c, \ldots$. There is a binary operation called *multiplication* for which, say, $ab = c$ is an element of the set, closed under successive multiplication. For $D = 1$, successive transformations of the form (1.14) have this property. The properties of groups will be discussed later in this book [see also Wigner, Jones], but here we wish to emphasize the meaning of the physical content of the result we have obtained through only the assumption, borne out in many experiments, that *light velocity measured in any inertial laboratory is the same.*

One can understand the central concept involved by considering an application of Einstein's result to the following experiments [Born]. Suppose the existence of two laboratories, called F_1 and F_2, capable of emitting and receiving light signals at any points in each laboratory, as we have considered in the beginning of this section. In the basic experiment, two signals are emitted from F. In F', they are detected at different times, for which the interval is changed. The times of emission are noted according to a clock in frame F, and the times of reception are noted on a clock in frame F'. *If the clock in frame F and the clock in F' are not identical, there is no way to compare the two intervals.* The statement of the experiment is meaningful only if there is a *universal time*, on all clocks intrinsic to any inertial frame.

We may identify this universal time with that postulated by Newton in Principia Mathematica, which we shall call τ to distinguish it from what we shall refer to as the Einstein time t, the outcome of a *measurement*, corresponding to the detection of the time signal in a given laboratory and recording its value as seen on the universal clock. The Einstein time t is therefore an *observable*, just as the spatial position of a particle, or its momentum.

As we shall see, it is the Einstein time t which enters Maxwell's description of elecromagnetism.

It is this notion of an absolute universal time which forms the main basis for the relativistic dynamics to be discussed in this book.

Chapter 2

Classical Relativistic Dynamics

2.1 Lorentz Group and Covariance

In the previous chapter, we have shown that Einstein's special relativity can be derived directly from the statement that the velocity of light measured in any inertial frame (a laboratory with measuring devices) must be the same. Since, in any frame, measured intervals of distance and time undergo transformations to $\Delta x'$ and $\Delta t'$, which in general, are different from the intervals Δx and Δt selected in the signal emitting laboratory. The requirement that the ratio for an emitted light signal be the same is far from trivial. The derivation that was given made use of mathematics only on a very elementary level (linear algebra).

To proceed, however, we will need some mathematical tools on a little higher level, which we explain in the following. As we have seen in the previous chapter, an essential requirement was that a succession of two transformations of this type must have the same form as a single transformation. This is a property of a *group*, a set of elements, as we pointed out towards the end of the previous chapter, which we may denote by $a, b, c, \ldots$, for which a binary relation ab, called *multiplication* is defined, with the property that $ab = c$, where c is another element of the set, for any a, b in the set. A group furthermore has the property that there is an identity element e satisfying $ea = ae = a$; multiplication is associative,*i.e.*, $a(bc) = (ab)c$ (but not necessarily commutative). Each element has an inverse a^{-1} for which $a^{-1}a = aa^{-1} = e$, and the set $a, b, c, \ldots$ is closed; any sequence of multiplications remains in the set. Simple examples are the real numbers under ordinary multiplication excluding zero, or the set of real numbers with ordinary addition as the binary operation with zero as the unit element. These four axioms (multiplication, associativity, closure and the existence of an inverse) form the basis of the theory of groups [Wigner (1931), Jones (1990)].

This mathematical structure has an obvious application to physical actions, which we may call *transformations*, such as pushing an object on a table, turning a sphere, or, to demonstrate noncommutativity, the discrete rotations of a blackboard eraser. The sequence of operations used in the previous chapter was the successive emission and detection of light signals by inertial laboratories.

We may represent the linear transformations (1.25) as (we suppress the Δ symbol in what follows, understanding all the variables as intervals):

$$x^{\mu'} = \Lambda^{\mu}{}_{\nu} x^{\nu}, \qquad (2.1)$$

where there is a sum (Einstein convention) over the indices μ, ν over $0, 1$, and the matrix $\Lambda^{\mu}{}_{\nu}$ consists of the coefficients entering in (1.25). We see that if we multiply $x^{0'}$ by -1 and $x^{1'}$ by $+1$, and call the result $x_{\mu}{}'$, then

$$x_{\mu}{}' x^{\mu'} = x_{\mu} x^{\mu} \qquad (2.2)$$

corresponds to the invariance requirement (1.4). The relation of x_{μ} to x^{μ} can be written as a matrix multiplication, *i.e.*, as $x_{\mu} = \eta_{\mu\nu} x^{\nu}$, where $\eta_{\mu\nu}$ is diagonal with $(-1, 1)$ as the diagonal elements. The matrix $\eta_{\mu\nu}$ is called a *metric* (since it provides a means for fixing a length), and serves to raise and lower indices; x^{μ} is called a *vector*, describing a direction and magnitude in the four dimensional spacetime.

The discussion above can be generalized to include *rotations*, for which the relation (2.2) includes a sum over space indices with additional elements $+1, +1$ in the diagonal metric for the y and z components; the spacial parts of the sum satisfy the requirement (2.2) since spacial rotations preserve the sum of squares of the spacial components (radius squared). With this, we see that the requirement

(2.2) is satisfied if

$$\Lambda^{\mu}{}_{\nu}\Lambda^{\nu}{}_{\lambda} = \eta^{\mu}{}_{\lambda}, \qquad (2.3)$$

where $\eta^{\mu}{}_{\lambda}$ is the 4×4 unit matrix, obtained by multiplying $\eta_{\nu\lambda}$ by $\eta^{\mu\nu}$, the same matrix as $\eta_{\mu\nu}$.

We recognize that the *vector x^{μ} transforms* under Lorentz transformations represented by $\Lambda^{\mu}{}_{\nu}$.

Furthermore, our requirement that successive transforations have the same form (from x' to x'') is then transcribed as

$$\Lambda^{\mu}{}_{\nu}\Lambda'^{\nu}{}_{\lambda} = \Lambda''^{\mu}{}_{\lambda}, \qquad (2.4)$$

so that we see that the set of Lorentz transformations $\{\Lambda^{\mu}{}_{\nu}\}$ form a *group*, which is called the *Lorentz group*.

Vectors of the type x^{μ} (or some arbitrary vector v^{μ}) are also called *tensors*, quantities that transform linearly under the action of a group. Such a tensor is called first rank. A quantity $w^{\mu\nu}$ transforming like $v^{\mu}u^{\nu}$ is a tensor of second rank; we shall have much to say about such objects in the sequel.

It is generally asserted that in every inertial frame the laws of physics should take the same form. This is the notion of Galileo's ship [Galileo (1632)], where inside a boat floating along driven by the wind at a constant speed, butterflies fly as in a garden at rest, a ball can be thrown in a path as if the floor were at rest, and smoke from a pipe curls upward as if the air were at rest. No sense of motion is perceived by the passengers unless they look out of the window. One concludes that the laws governing observed dynamical phenomena should be *invariant* under transformations from one inertial frame to another, *i.e.* observations of a dynamical process carried out in any inertial frame should differ, in the relativistic case, only, as observed, by Lorentz transformation of the observables. One therefore infers that physical laws should be formulated in a way which have this property of *Lorentz invariance*.

2.2 Mass and Momentum

In the nonrelativistic theory, Newton's laws of motion [Newton (1687)], specifies that the momentum ($\mathbf{p} \equiv m\mathbf{v}$) of a particle changes only under application of a force, for which $\frac{d}{dt}\mathbf{p} = \mathbf{f}$, where $\mathbf{f}$ is the force. In a similar way, in order to formulate dynamical laws which can characterize the motion of a particle in a relativistic framework, we must define a momentum *four vector* which reduces to the form ($\mathbf{p} = m\mathbf{v}$) in the nonrelativistic limit. Since Maxwell's electromagnetism [Maxwell (1865)] is relativistically covariant [Einstein (1905)], we may start by deriving one of the most striking results of special relativity, the relation between mass and energy ($E = mc^2$) from the Maxwell theory, supplemented by Poynting's theorem [Poynting (1884)], that a light wave falling on a body exerts a pressure on it. Following the argument of Max Born (1962), one concludes that the momentum transfered to an absorbing surface due to a short flash of light is E/c, confirmed experimentally by Lebedev (1890) and Nichols and Hull (1901). The same pressure would be experienced by an emitter of light (recoil).

Consider a long tube of length ℓ with two bodies A and B of mass M, one at each end. Suppose A has an excess of energy in the form of heat, which can then be sent to B by radiation in a short pulse. A then experiences a recoil E/c, providing the body with a speed v such that $Mv = E/c$. This motion continues until the pulse strikes B, and then is brought to a stop by the absorption. The displacement during the time of travel, approximately $t = \ell/c$, $x = vt$, is then

$$x = \frac{E\ell}{Mc^2}. \tag{2.5}$$

As Born (1962) points out, this situation is untenable, since the box has moved with no external force. We must therefore assign an additional mass to A to account for its additional energy initially. The additional momentum $-mc = -m\frac{\ell}{t}$ must compensate for the recoil $Mv = M\frac{x}{t}$, so that

$$Mx - m\ell = 0, \tag{2.6}$$

so that, to compensate the recoil displacement, we must have

$$x = \frac{m\ell}{M} = \frac{E\ell}{Mc^2}. \tag{2.7}$$

Then, independently of M and t, one obtains

$$m = \frac{E}{c^2}, \tag{2.8}$$

the mass equivalent of the energy of the radiation. Following Born's argument, we have arrived at the well-known result $E = mc^2$.

We now turn to Einstein's argument [Einstein (1922)] defining the four-momentum, *i.e.*, that

$$p_\mu = \left(\mathbf{p}, \frac{E}{c} \right) \tag{2.9}$$

is the only four-vector function of momentum and energy for which the result (2.8) emerges at rest, and *defines* the dynamically measured mass m by

$$-\eta^{\mu\nu} p_\mu p_\nu = m^2 c^2. \tag{2.10}$$

As Einstein remarked [Einstein (1922)], one may define the invariant ds on the particle path ($dx^0 \equiv cdt$)

$$-c^2 ds^2 = (dx^1)^2 + (dx^2)^2 + (dx^3)^2 - (dx^0)^2 \tag{2.11}$$

so that, for

$$u^\mu \equiv \frac{dx^\mu}{ds}, \tag{2.12}$$

we obtain

$$u_\mu u^\mu = -c^2. \tag{2.13}$$

As for dx^μ, the quantity u^μ is a four-vector, and (2.13) is invariant. Moreover, from (2.11) ($v = |\mathbf{v}|$),

$$\frac{ds}{dt} = \pm \sqrt{1 - \left(\frac{v}{c} \right)^2}, \tag{2.14}$$

where $\mathbf{v} = \frac{d\mathbf{x}}{dt}$, the velocity observed of a particle moving with frame F as seen in our laboratory frame F'. For v small compared to c, and t positive, ds is well approximated by dt, and u^μ becomes approximately $(\mathbf{v}, c)$.

Since, however, the intervals $d\mathbf{x}$, dt, for the motion of a particle, are *dynamical*, the proper time ds is not a good parameter for use in the equations describing the motion. Stueckelberg (1941), as we discuss in the next section, argued that in this dynamical context, giving the example of *classical pair annihilation*, one must introduce an invariant parameter, which he called τ, to effectively describe the motion. Since it is recognized that dx, dt are dynamical, it follows that $\mathbf{p}$ and E are also dynamical variables, and therefore the mass m in (2.10) is also a dynamical variable. In our discussion of the four momentum, we shall follow the approach of Einstein, but take into account that the mass m of the particle is variable.

Let us define the four vector

$$p^{\mu} = M\dot{x}^{\mu}, \tag{2.15}$$

where M is a constant dimensional parameter. Then, by (2.11),

$$p^{\mu}p_{\mu} = -M^2 c^2 \frac{ds^2}{d\tau^2}. \tag{2.16}$$

By (2.10), however, we must have[1]

$$\frac{ds^2}{d\tau^2} = \frac{m^2}{M^2}, \tag{2.17}$$

where m is the (variable) measured mass squared of the particle, so that (2.10) is satisfied. With fourth component $p_0 = E/c$, we would have,

$$p_{\mu}p^{\mu} = -m^2 c^2. \tag{2.18}$$

At rest, for $\mathbf{p} = 0$, this would result in

$$\frac{E^2}{c^2} = m^2 c^2, \tag{2.19}$$

or, as required, $E = mc^2$. With this, following Einstein's argument, modified to take into account the possible variation of the mass m with τ, we have identified the energy momentum four vector (2.15).

---

[1]We shall see in the discussion of Stueckelberg's theory [Stueckelberg (1941)] below, that this relation follows rigorously in a consistent relativistically dynamical framework.

The coordinates $\{x^\mu\}$ and momenta $\{p_\mu\}$ provide us with an eight dimensional phase space. In the six dimensional phase space of non-relativistic mechanics, the *state* of a particle, represented by a point in this space, moves according to the equations of motion with the Newtonian time t. This description is not consistent with relativity, since, seen from a moving frame, the time and space variables would become mixed. The state of a particle, represented by a point in the eight dimensional phase space, providing a framework consistent with relativity, evolves with an invariant parameter τ which plays the role of the Newtonian time.

2.3 Pair Annihilation in Classical Mechanics

Stueckelberg (1941), imagined the motion of a free particle in spacetime on a rectilinear system of coordinates with axes t and x as following a linear path with x proportional to t with, as usual, proportionality constant v. This line, however, should not be thought of as simply a plot of the *a posteriori* solution of a linear differential equation, but as the track of the *motion of an event* x^μ on the manifold of spacetime, determined by dynamical equations. In Steuckelberg's picture, this event encounters an interaction at some point P, and is deflected in its development to a motion in the negative direction of the t axis, and then proceeds freely. The resulting picture of this history is that of two straight lines converging (in the sense of positive t) to the point P. Stueckelberg interpreted the track of the event moving backwards in t (with Dirac [Dirac (1930), (1932), (1938)]) as an *antiparticle*, thus representing particle-antiparticle annihilation in classical mechanics (Fig. 1).

He furthermore pointed out that the resulting orbit cannot be parametrized by t, since it would, in that parametrization, be double valued. He therefore introduced a new (invariant) parameter τ, providing the possibility of writing equations for the motion of an event described by $x^\mu(\tau)$. It is this very profound notion of the evolution of a particle as the motion of an event on the manifold of spacetime which forms the basis for the content of this book.

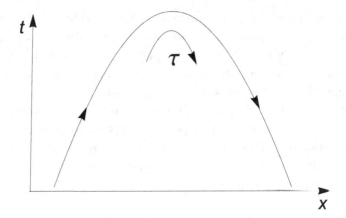

Fig. 1.   World line for pair annihilation in classical mechanics. The segment going in the negative time direction is intepreted as an antiparticle going forward in time. (Courtesy Springer, Dordrecht).

Based on ideas of Fock (1937), Steuckelberg then postulated a dynamical formulation which reduces, in the non-relativistic limit, to the usual non-relativistic form, but is completely relativistically covariant. He assumed the existence of a Hamilton-Lagrange theory (*e.g.* [Goldstein (1951)]) with a form analogous to that of the non-relativistic theory. In the next section, we formulate Stueckelberg's assertion starting from basic principles.

2.4   Relativistic Hamilton-Lagrange Dynamics

In this section, we formulate a relativistic form of Hamilton-Lagrange dynamics [Goldstein (1951)], which forms the basis of Stueckelberg's covariant classical (and quantum) dynamics.

As we have defined in (2.15), $p_\mu \equiv M u_\mu$, which is constant in the absence of forces. A force acting on the particle will change the momentum (as occurs in the example of pair anihilation discussed above); we may describe such a force as ($\dot{p}_\mu \equiv \frac{d}{d\tau} p_\mu$, a notation we will use throughout):

$$\dot{p}_\mu = M \dot{u}_\mu = \eta_{\mu\nu} M \ddot{x}^\nu, \qquad (2.18)$$

a generalization of Newton's law $\mathbf{F} = m\mathbf{a}$. This generalized force multiplied by the velocity u^μ gives the rate of transfer of energy (mass) to the system

$$\dot{p}_\mu \dot{x}^\mu = M \frac{1}{2} \frac{d}{d\tau} \dot{x}^\mu \dot{x}_\mu, \tag{2.19}$$

generalized to the eight dimensional space.

In this way, we have identified the change in momentum with a force according to

$$F_\mu = \dot{p}_\mu; \tag{2.20}$$

one can therefore think of the relation

$$F_\mu - \dot{p}_\mu = 0 \tag{2.21}$$

as a statement of equilibrium. For a small displacement from equilibrium, we would have a generalized *D'Alembert principle* [Goldstein (1951)]

$$(F_\mu - \dot{x}_\mu)\delta x^\mu = 0. \tag{2.22}$$

Let us consider a small motion along the particle trajectory as the variation. It will be convenient to introduce a generalized coordinate s_μ along the path, so that

$$\delta x^\mu = \frac{\partial x^\mu}{\partial s_\nu} ds_\nu. \tag{2.23}$$

For the second term of (2.22), we then have

$$M\ddot{x}_\mu \frac{\partial x^\mu}{\partial s_\nu} = \frac{d}{d\tau}\left(M\dot{x}_\mu \frac{\partial x^\mu}{\partial s_\nu} \right) - M\dot{x}_\mu \frac{\partial \dot{x}^\mu}{\partial s_\nu} \tag{2.24}$$

Since $x^\mu = x^\mu(s)$ (we assume no explicit τ dependence in this transformation), it follows that

$$\dot{x}_\mu = \frac{\partial x^\mu}{\partial s_\nu} \dot{s}_\nu, \tag{2.25}$$

so that

$$\frac{\partial x^\mu}{\partial s_\nu} = \frac{\partial \dot{x}^\mu}{\partial \dot{s}_\nu}. \tag{2.26}$$

Then, for (2.24),

$$M\ddot{x}_\mu \frac{\partial x^\mu}{\partial s_\nu} = \frac{d}{d\tau}\left(M\dot{x}_\mu \frac{\partial \dot{x}^\mu}{\partial \dot{s}_\nu}\right) - M\dot{x}_\mu \frac{\partial \dot{x}^\mu}{\partial s_\nu} \qquad (2.27)$$

Let us call

$$T = \frac{1}{2}M\dot{x}_\mu \dot{x}^\mu, \qquad (2.28)$$

the *kinetic energy*, in analogy to the non-relativistic theory. Then, (2.22) becomes

$$F_\mu \frac{\partial x^\mu}{\partial s_\nu}\delta s^\nu = \left\{\frac{d}{d\tau}\frac{\partial T}{\partial \dot{s}_\nu} - \frac{\partial T}{\partial s_\nu}\right\}\delta s^\nu. \qquad (2.29)$$

For a potential type model, with potential energy (mass) $V(x)$,

$$F_\mu \frac{\partial x^\mu}{\partial s_\nu} = -\frac{\partial V}{\partial x^\mu}\frac{\partial x^\mu}{\partial s_\nu} = -\frac{\partial V}{\partial s_\nu}, \qquad (2.30)$$

so that

$$\left\{\frac{d}{d\tau}\frac{\partial T}{\partial \dot{s}_\nu} - \frac{\partial(T-V)}{\partial s_\nu}\right\}\right\} = 0. \qquad (2.31)$$

For potential models, V generally does not depend on $\dot{s}_\nu{}^2$ so we may write (2.31) as

$$\frac{d}{d\tau}\frac{\partial L}{\partial \dot{s}_\nu} - \frac{\partial L}{\partial s_\nu} = 0, \qquad (2.32)$$

where

$$L = T - V, \qquad (2.33)$$

is called the *Lagrangian*, the basis for the Hamilton-Lagrange mechanics. [Goldstein (1951)] In terms of the original variables x^μ, we may write this formula as

$$\frac{d}{d\tau}\frac{\partial L}{\partial \dot{x}^\mu} - \frac{\partial L}{\partial x^\mu} = 0. \qquad (2.34)$$

We now turn to the definition of the Hamiltonian and the associated canonical variables. Writing

$$L = \frac{1}{2}\dot{x}_\mu \dot{x}^\mu - V(x), \qquad (2.35)$$

---

[2]In the presence of electromagnetic fields, the interactions may be velocity dependent, as we shall discuss later.

we have

$$\frac{\partial L}{\partial \dot{x}^\mu} = M \dot{x}_\mu = p_\mu, \qquad (2.36)$$

which may be taken as a general definition of p_μ. We may now write a *conservation theorem*. Consider the relation

$$\frac{dL}{d\tau} = \frac{\partial L}{\partial x^\mu} \frac{dx^\mu}{d\tau} + \frac{\partial L}{\partial \dot{x}^\mu} \frac{d\dot{x}^\mu}{d\tau}. \qquad (2.37)$$

Since, however,

$$\frac{\partial L}{\partial x^\mu} = \frac{d}{d\tau} \frac{\partial L}{\partial \dot{x}^\mu}, \qquad (2.38)$$

we see that

$$\begin{aligned}
\frac{dL}{d\tau} &= \frac{d}{d\tau} \frac{\partial L}{\partial \dot{x}^\mu} + \frac{\partial L}{\partial \dot{x}^\mu} \frac{d\dot{x}^\mu}{d\tau} \\
&= \frac{d}{d\tau}\left(\frac{\partial L}{\partial \dot{x}^\mu} \dot{x}^\mu \right).
\end{aligned} \qquad (2.39)$$

It then follows that

$$\frac{d}{d\tau}\left(L - \frac{\partial L}{\partial \dot{x}^\mu} d\dot{x}^\mu \right) = 0, \qquad (2.40)$$

or

$$\frac{dK}{d\tau} = 0, \qquad (2.41)$$

where

$$K \equiv \frac{\partial L}{\partial \dot{x}^\mu} d\dot{x}^\mu - L. \qquad (2.42)$$

Using the relation (2.36), we can write (2.42) as

$$K = \dot{x}^\mu p_\mu - L, \qquad (2.43)$$

a constant of the motion, called the *Hamiltonian* for relativistic dynamics. With L of the form (2.35), (2.43) can be written as

$$K = \frac{1}{2} \dot{x}_\mu \dot{x}^\mu + V(x) \qquad (2.44)$$

or

$$K = \frac{p^\mu p_\mu}{2M} + V, \qquad (2.45)$$

the form assumed by Stueckelberg (1941) to formulate his relativistic dynamics.

The corresponding *canonical* equations of motion are, in general,

$$\dot{x}^\mu = \frac{\partial K}{\partial p_\mu}$$
$$\dot{p}_\mu = -\frac{\partial K}{\partial x^\mu},$$

(2.46)

the first giving, in this case,

$$\dot{x}^\mu = \frac{p^\mu}{M}$$

(2.47)

and the second, the conservative force relation,

$$\dot{p}_\mu = -\frac{\partial V}{\partial x^\mu}.$$

(2.48)

As remarked above, it follows from (2.47) that

$$\frac{dx^\mu dx_\mu}{d\tau^2} = \frac{p^\mu p_\mu}{M^2},$$

(2.49)

or, by (2.11) and (2.18), we obtain (2.17)

$$\frac{ds^2}{d\tau^2} = \frac{m^2}{M^2}.$$

(2.17)

If the dynamical particle mass m is equal to the dimensional parameter M, we shall say that the particle is "on-shell" (this "shell" is the surface in momentum space for which $p_\mu p^\mu = -M^2 c^2$).

With (2.46), we may define the form of the covariant *Poisson bracket* [Goldstein (1951)]. To do this, we compute the total τ derivative of a function $F(x, p)$,

$$\frac{dF(x,p)}{d\tau} = \frac{\partial F(x,p)}{\partial x^\mu}\dot{x}^\mu + \frac{\partial F(x,p)}{\partial p_\mu}\dot{p}_\mu$$

$$= \frac{\partial F(x,p)}{\partial x^\mu}\frac{\partial K}{\partial p_\mu} - \frac{\partial F(x,p)}{\partial p_\mu}\frac{\partial K}{\partial x^\mu}$$

$$\equiv [F, K]_{PB},$$

(2.50)

the Poisson bracket of F and K. We therefore have

$$\frac{\partial F(x,p)}{\partial x^\mu} = [F, p_\mu]_{PB},$$

$$\frac{\partial F(x,p)}{\partial p_\mu} = -[F, x^\mu]_{PB}.$$

(2.51)

The variables x^μ, p_ν therefore, as in the non-relativistic theory, act as derivatives under the Poisson bracket. The Poisson bracket itself is often called a *derivation*, and satisfies the Jacobi identity [Goldstein (1951)]

$$[A, [B, C]_{PB}]_{PB} + [B, [C, A]_{PB}]_{PB} + [C, [A, B]_{PB}]_{PB} = 0. \quad (2.52)$$

We furthermore have the *canonical Poisson bracket relations*

$$[x^\mu, p_\nu]_{PB} = \eta^\mu{}_\nu, \quad (2.53)$$

where $\eta^\mu{}_\nu$ is the diagonal matrix $(1, 1, 1, 1)$ we used in (2.3).

In terms of these Poisson bracket relations, we remark here that the antisymmetric tensor

$$M_{\mu\nu} = x_\mu p_\nu - x_\nu p_\mu, \quad (2.54)$$

under Poisson bracket with x^λ, *generates* an infinitesimal Lorentz transformation:

$$[M_{01}, x^0]_{PB} = -x_1 \quad (2.55)$$

and

$$[M_{01}, x^1]_{PB} = -x_0 \quad (2.56)$$

in agreement with the infinitesimal form of (1.46) (with small coefficient v). In the same way, the (i, j) spatial components of $M_{\mu\nu}$ generate infinitesimal rotations on the space vector (x_1, x_2, x_3). We therefore call $M_{\mu\nu}$ the *relativistic angular momentum tensor*. We shall see in the next chapter how these relations play an important role in the relativistic quantum theory.

The existence of a set of canonical variables and a Hamiltonian provide us with a framework for canonical dynamics.[3]

---

[3]Following Dirac (1947), this forms the basis for the quantization of the theory, as we shall see in the next chapter.

The development of this section has been so far concerned with the dynamics of a single particle. In order to be able to discuss many body systems, [Horwitz and Piron (1973), Horwitz (1973)] made a far reaching and fundamental assumption; that *the parameter τ is universal*, that it has the same role in the correlation and description of many body systems as Newtonian time for the non-relativistic world (to be called SHP), making possible the development of relativistic statistical mechanics [Schieve (2009)]. This basic assumption underlies all of the discussion that follows.

2.5 The Classical Two Body System

For the two body system, the Hamiltonian (2.45) can be generalized additively[4] to

$$K = \frac{p_1{}^\mu p_{1\mu}}{2M_1} + \frac{p_2{}^\mu p_{2\mu}}{2M_2} + V(x_1 - x_2), \qquad (2.57)$$

where, for a model with Lorentz invariance and translation invariance (Poincaré invariance), we take V to be a function of $(x_1{}^\mu - x_2{}^\mu)(x_{1\mu} - x_{2\mu})$. Both coordinates $x_1{}^\mu$ and $x_2{}^\mu$ are parametrized in their motions by the *same* universal τ, ensuring the correlation that enables us to establish an interaction.[5]

As a justification for assuming the form (2.54), we see from (2.50) that the τ derivative of a product of a function of $x_1{}^\mu$ and a function of $x_2{}^\mu$ is generated by the Poisson bracket with the form (2.54).

As for the non-relativistic two body problem, we may define the relative and "center of mass" coordinates,

$$P^\mu = p_1{}^\mu + p_2{}^\mu \quad p^\mu = \frac{M_1 p_2{}^\mu - M_2 p_1{}^\mu}{M_1 + M_2}$$

$$X^\mu = \frac{M_1 x_1{}^\mu + M_2 x_2{}^\mu}{M_1 + M_2} \quad x^\mu = x_2{}^\mu - x_2{}^\mu. \qquad (2.58)$$

---

[4]The additivity of the kinetic terms corresponds to the additivity of the mass for two noninteracting subsystems.

[5]I am grateful to E.C.G. Sudarshan (1998) for a discussion of this point.

The two body Hamiltonian then takes the form

$$K = \frac{P^\mu P_\mu}{2M_T} + \frac{p^\mu p_\mu}{2M_{rel}} + V(x)$$

$$= K_T + K_{rel},$$

(2.59)

where

$$M_T = M_1 + M_2 \quad M_{rel} = \frac{M_1 M_2}{M_1 + M_2}$$

(2.60)

and

$$K_{rel} = \frac{p_\mu p^\mu}{2M_{rel}} + V(x).$$

(2.61)

To prove the equivalence between (2.57) and (2.59), we use

$$\frac{M_2 P_\mu}{M_1 + M_2} = \frac{M_2 p_{1\mu}}{M_1 + M_2} + \frac{M_2 p_{2\mu}}{M_1 + M_2}$$

(2.62)

so that, after some cancellations, with (2.58),

$$p_\mu - \frac{M_2 P_\mu}{M_1 + M_2} = -p_{2\mu}.$$

(2.63)

In a similar way,

$$\frac{M_1 P_\mu}{M_1 + M_2} = \frac{M_1 p_{1\mu}}{M_1 + M_2} + \frac{M_1 p_{2\mu}}{M_1 + M_2}$$

(2.64)

so that

$$p + \frac{M_1 P_\mu}{M_1 + M_2} = p_{1\mu}.$$

(2.65)

Then, for (2.54), we have

$$\frac{p_1{}^\mu p_{1\mu}}{2M_1} = \frac{p^\mu p_\mu}{2M_1} + \frac{p_\mu P^\mu}{M_1 + M_2} + \frac{1}{2} \frac{M_1 P_\mu P^\mu}{(M_1 + M_2)^2}$$

(2.66)

and

$$\frac{p_2{}^\mu p_{2\mu}}{2M_2} = \frac{p^\mu p_\mu}{2M_2} - \frac{p_\mu P^\mu}{M_1 + M_2} + \frac{1}{2} \frac{M_2 P_\mu P^\mu}{(M_1 + M_2)^2}.$$

(2.67)

In the sum of these two expressions, the second terms on the right cancel, and we obtain (with (2.60)) the formula (2.59).

Since, in the absence of external forces, the total energy-momentum P_μ is conserved, K_T is conserved, and therefore K_{rel} is a constant of the motion as well.[6]

We now shown that the equations of motion for the *relative motion* are generated by canonical relations of the same form as for the one body problem, so that the relative motion may be considered as a reduction of the two body problem to a one body (reduced) motion.

We first compute the τ derivative of the relative coordinate x^μ:

$$\dot{x}_1^\mu - \dot{x}_2^\mu = \frac{p_1{}^\mu}{M_1} - \frac{p_2{}^\mu}{M_2}$$

$$= \frac{M_2 p_1{}^\mu - M_1 p_2{}^\mu}{M_1 M_2} \tag{2.68}$$

$$= \frac{p^\mu}{M_{rel}},$$

so that

$$\dot{x}^\mu = \frac{p^\mu}{M_{rel}}. \tag{2.69}$$

Furthermore,

$$\dot{p}_\mu = \frac{M_2 \dot{p}_{1\mu} - M_1 \dot{p}_{2\mu}}{M_1 + M_2}$$

$$= \frac{1}{M_1 + M_2}\left\{ -M_2 \frac{\partial V}{\partial x_1{}^\mu} + M_1 \frac{\partial V}{\partial x_2{}^\mu} \right\} \tag{2.70}$$

$$= -\frac{\partial V}{\partial x^\mu}.$$

Finally,

$$\dot{P}_\mu = -\frac{\partial V}{\partial x_1{}^\mu} - \frac{\partial V}{\partial x_2{}^\mu} = 0, \tag{2.71}$$

---

[6]Note that

$$(p_1{}^\mu + p_2{}^\mu)(p_{1\mu} + p_{2\mu}) \equiv s$$

is often called the Mandelstam [Chew (1966)] invariant corresponding to the total mass squared in the input scattering channel; we shall refer to this variable later.

as expected, and

$$\dot{X}^\mu = \frac{M_1 \dot{x}_1^\mu + M_2 \dot{x}_2^\mu}{M_1 + M_2}$$
$$= \frac{p_1{}^\mu + p_2{}^\mu}{M_1 + M_2} = \frac{P^\mu}{M_T}. \tag{2.72}$$

We have therefore shown that the canonical equations derived from K_T and K_{rel}, as if these describe two independent systems, are equivalent to the canonical equations for the two body system obtained from (2.57).

We furthermore remark that, in a similar way, the sum of the angular momenta of the two subsystems is equal to the angular momentum of the center of mass system and the angular momentum of the system with relative coordinates. This is shown most easily by substituting the definitions (2.58) into

$$M_{T\mu\nu} = X_\mu P_\nu - X_\nu P_\mu + x_\mu p_\nu - x_\nu p_\mu \tag{2.73}$$

to obtain, after some cancellations,

$$M_{T\mu\nu} = x_{1\mu} p_{1\nu} - x_{1\nu} p_{1\mu} + x_{2\mu} p_{2\nu} - x_{2\nu} p_{2\mu}, \tag{2.74}$$

the sum of the generators of the two subsystems. Since the Poisson brackets of both of the generators with the Hamiltonian, for a scalar potential, vanish, they are both conserved quantities.

It was stated in the original paper of Horwitz and Piron [Horwitz (1973)] and in [Fanchi (1993)] that the orbits and momenta of the relative motion in the presence of an invariant potential are restricted to a two dimensional plane. In the following, we give a detailed proof of this assertion. To do this we use two fundamental identities. The first is

$$M^{\mu\nu} x^\lambda + M^{\lambda\mu} x^\nu + M^{\nu\lambda} x^\mu = 0 \tag{2.75}$$

explicitly for indices $\{123, 012, 023, 013\}$, and the second,

$$M_{\mu\nu} M_{\rho\lambda} \epsilon^{\mu\nu\rho\lambda} = 0, \tag{2.76}$$

where $\epsilon^{\mu\nu\rho\lambda}$ is the totally antisymmetric tensor with entries ± 1.

The first of these relations follows explicitly from the definition of $M^{\mu\nu}$:

$$(x^\mu p^\nu - x^\nu p^\mu)x^\lambda + (x^\lambda p^\mu - x^\mu p^\lambda)x^\nu$$
$$+ (x^\nu p^\lambda - x^\lambda p^\nu)x^\mu = 0 \tag{2.77}$$

by cancellation of terms.

The second follows directly due to the symmetry in spacetime coordinates (or energy-momentum):

$$M_{\mu\nu}M_{\rho\lambda}\epsilon^{\mu\nu\rho\lambda} = \epsilon^{\mu\nu\rho\lambda}(x^\mu p^\nu - x^\nu p^\mu)(x^\rho p^\lambda - x^\lambda p^\rho) = 0 \tag{2.78}$$

since each term contains products of symmetric quadratic forms with the totally antisymmetric tensor $\epsilon^{\mu\nu\rho\lambda}$.

The first provides four homogeneous equations for x^0, x^1, x^2 and x^3. In matrix form, these are

$$\begin{pmatrix} 0 & M^{23} & M^{31} & M^{12} \\ M^{12} & M^{20} & M^{01} & 0 \\ M^{23} & 0 & M^{30} & M^{02} \\ M^{13} & M^{30} & 0 & M^{01} \end{pmatrix} \begin{pmatrix} x^0 \\ x^1 \\ x^2 \\ x^3 \end{pmatrix} = 0. \tag{2.79}$$

For non-zero solutions, the determinant must vanish, so that there must be a linear dependence between two rows. We now show that the condition (2.76) leads to a constraint that the motion is restricted to a plane, *i.e.* there are just two independent degrees of freedom.

Multiplying out the terms (on the pairs μ, ν) in (2.76), after several cancellations, we are left with

$$M^{01}M^{23} + M^{02}M^{31} + M^{03}M^{12} = 0 \tag{2.80}$$

Multiplying the first row in (2.79) by M^{01}, the second by $-M^{31}$ and fourth by $-M^{12}$, the sum of these three rows add to zero, thus providing a proof of linear dependence, leaving the relations (2.79) with solutions on a plane.

It is an interesting property of the relations (2.75) that multiplying by the dual form $\epsilon_{\alpha\beta\mu\nu}M^{\alpha\beta}$, the first term vanishes by (2.76). The result takes a simple form if we call

$$\epsilon_{\alpha\beta\mu\nu}M^{\alpha\beta}x^\nu \equiv N_\mu. \tag{2.81}$$

One then obtains

$$\hat{M}_{\mu\nu}M^{\lambda\mu}x^{\nu} = 0, \tag{2.82}$$

where

$$\hat{M}_{\mu\nu} = \epsilon_{\alpha\beta\mu\nu}M^{\alpha\beta}. \tag{2.83}$$

From (2.82), we see that

$$M^{\lambda\mu}N_{\mu} = 0. \tag{2.84}$$

The Lorentz transform of a non-vanishing vector cannot be zero, and therefore

$$N_{\mu} = 0. \tag{2.85}$$

Therefore,

$$N_{\mu} \rightleftharpoons \hat{M}_{\mu\nu}x^{\mu} = 0 \tag{2.86}$$

which can be written as

$$\begin{pmatrix} 0 & M^{23} & -M^{13} & M^{12} \\ -M^{23} & 0 & -M^{30} & M^{20} \\ M^{13} & M^{30} & 0 & M^{01} \\ -M^{12} & -M^{20} & -M^{01} & 0 \end{pmatrix} \begin{pmatrix} x^0 \\ x^1 \\ x^2 \\ x^3 \end{pmatrix} = 0. \tag{2.87}$$

or

$$\begin{aligned} M^{23}x^1 + M^{31}x^2 + M^{12}x^3 &= 0 \\ -M^{23}x^0 - M^{30}x^2 + M^{20}x^3 &= 0 \\ M^{13}x^0 + M^{30}x^1 + M^{01}x^3 &= 0 \\ -M^{12}x^0 - M^{20}x^1 - M^{01}x^2 &= 0. \end{aligned} \tag{2.89}$$

The set of equations for x^0, x^1, x^2, x^3 obained in this way are identical to those obtained from (2.79). The projection removing the first term of (2.75) implicitly eliminates the information contained in the condition (2.76) required to reduce the space of solutions to two dimensions; what remains is a demonstration of the consistency of our identities.

This completes our discussion of how the identities (2.75) and (2.76) restrict the motion of a particle in a central potential to a plane. A similar procedure applies for the four momenta.[7]

---

[7]Although this result has been stated in the references above, the demonstration given here has not, to our knowledge, appeared before this writing.

2.6 Some Examples

In this section, we give some examples of two-body systems with potential model interactions. In particular, we treat here the relativistic harmonic oscillator for classical relativistic mechanics. The quantum mechanical oscillator [Leutwyler (1977), Kim and Noz (1977), Horwitz (1973)] will be discussed in Chapter III. We shall also discuss the relativistic two body Kepler problem, where, in contrast to the result of Sommerfeld [ref], we show that there is no precession of the orbit [Horwitz and Piron (1973)].

2.6.1 *The Classial Relativistic Harmonic Oscillator*

The Hamiltonian for the relativistic harmonic oscillator is

$$K_{rel} = \frac{P_\mu P^\mu}{2M_T} + \frac{p_\mu p^\mu}{2M_{rel}} + k x^\mu x_\mu, \qquad (2.90)$$

where we have used the center of mass and relative coordinates and momenta as discussed before, and k is the "spring" constant. The variable $x^\mu = x_1{}^\mu - x_2{}^\mu$, generally spacelike in the nonrelativistic case, may run over values that are spacelike or timelike. Our model interaction is not limited by causal restrictions, but understood as an effective interaction of a covariant dynamics on the Minkowski space.

The equations of motion for the reduced motion, (2.69) and (2.71) then read

$$\dot{x}^\mu = -\frac{\partial K_{rel}}{\partial p_\mu} = -\frac{p^\mu}{M_{rel}} \qquad (2.91)$$

and therefore

$$\ddot{x}^\mu = -\frac{k}{M_{rel}}. \qquad (2.92)$$

The general solutions are of the form

$$x^\mu(\tau) = a^\mu \cos(\omega\tau) + b^\mu \sin(\omega\tau), \qquad (2.93)$$

where

$$\omega = \sqrt{\frac{k}{M_{rel}}}. \qquad (2.94)$$

If we choose a^μ and b^μ to be spacelike, then, with τ, the particle event (representing *relative motion* will trace out a curve around and exterior to the light cone; although both x^0 and $\mathbf{x}$ oscillate, it is always true that $x^0 < |\mathbf{x}|$, and therefore the trajectory cannot pass through the light cone.

However, if we choose a^μ and b^μ to be timelike, then, with τ, the relative motion will trace out a curve that necessarily intersects both the forward and backward light cones.[8]

2.6.2 *The Classical Relativistic Kepler Problem*

The relativistic Kepler problem [Horwitz (1973)] is defined by choosing the potential function

$$V(x) = -\frac{q^2}{\rho}, \tag{2.95}$$

where $x^\mu = x_1{}^\mu - x_2{}^\mu$ and

$$\rho = +\sqrt{(x_1{}^\mu - x_2{}^\mu)(x_{1\mu} - x_{2\mu})}. \tag{2.96}$$

This form reduces in the non-relativistic limit for which $c \to \infty$ to the nonrelativistic Coulomb potential for two particles of charge q. If we take initial conditions on x^μ to be spacelike (for which $|\mathbf{x}| > |t|$), we may choose a Lorentz frame for which at every τ; t, p^0 and M_{0j} vanish, and orient the invariant two dimensional plane of motion in the spatial subspace.[9] The Hamilton equations (2.46)

$$\frac{d\mathbf{p}}{d\tau} = -\nabla V \quad \frac{dp^0}{d\tau} = -\frac{\partial V}{\partial t}$$

$$\frac{d\mathbf{x}}{d\tau} = \frac{\mathbf{p}}{M_{rel}} \quad \frac{dt}{d\tau} = \frac{p^0}{M_{rel}} \tag{2.97}$$

---

[8]There are no causal bounds, since our model interaction is action at a distance.

[9]From the first of (2.89), we see that in this configuration, $\mathbf{M} \cdot \mathbf{x} = 0$, where $\mathbf{M} = (M^{23}, M^{31}, M^{12})$; for the plane (x, y), $\mathbf{M}$ would lie along the z axis. Furthermore, as we argue below, we may take $x^0 = 0$; then, in general, the three last equations of (2.89) display the intersections of the plane of motion with the coordinate planes.

and, since E is constant, the center of mass time is then

$$T = t_1 = t_2 = \frac{E}{M_T}\tau = \frac{p_1{}^0 + p_2{}^0}{M_1 + M_2} \tag{2.98}$$

The measured times associated with each particle are, in this configuration, the same as the measured time T of the center of mass of the system. These results are the same as for the non-relativistic system, and the solutions of the relativistic problem then coincide with a Lorentz transformation of the classical solutions with orbits parametrized by τ, making explicit the interpretation of τ as the time originally postulated by Newton (1687).

One sees, furthermore, that in this framework there is no precession of the orbit of the Kepler problem, in contrast to a result of Sommerfeld (see, for example, [Horwitz (1973)]) (the observed precesssion of Mercury around the Sun is explained in Einstein's general theory of relativity).

In the next chapter, we begin our discussion of the foundations of the quantized theory of relativistic dynamics and its applications.

Chapter 3

Quantum Relativistic Dynamics

3.1 Introduction to the Quantum Theory: History and Foundations

Early experiments measuring radiation from a "black body", for example, a metal box with a small hole such that light entering does not emerge, or a stack of razor blades end on with the same property, detected radiation as a function of temperature with frequency distribution that followed a formula of Wien for low temperatures and of Rayleigh for high temperatures. Max Planck discovered that the formula [Born (1920)]

$$B(\nu, T) = \frac{2h\nu^3}{c^2} \frac{1}{e^{\frac{h\nu}{k_B T}} - 1}, \tag{3.1}$$

in units of watts per steradian, per meter squared per Hertz (cycles per second), written assuming that the light comes in little packages (quanta) of energy $h\nu = \hbar\omega$, where $h \equiv 2\pi\hbar$ and $\nu \equiv \omega = \frac{\nu}{2\pi}$. Expanding (3.1) in power series, we see contributions from one quantum, two quanta, *etc.*, as a discrete approximation to the integral describing Wien's law. This result however, succeeded in describing both the high and low temperature properties of the black body radiation.

In this same period, around 1900, when it appeared that Maxwell's theory of radiation was "working itself out beautifully" [Heitler (1936)], it became evident that hot gasses had a frequency dependent radiation that showed strong peaks at certain values. Niels Bohr (1913) offered an explanation which involved atoms with

31

electrons in Kepler like orbits that jumped from one orbit to another, emitting a definite discrete frequency of radiation. From a relativistic point of view, de Broglie (1925) suggested that the electrons in these orbits were controlled by waves of wavelength $\lambda = \frac{h}{p}$ and $E = h\nu$ and the orbits were characterized by integral numbers of wavelengths, partly accounting for jumps, and partly explaining Planck's quanta.

Based on de Broglie's use of relativity and his construction of waves, the mass energy relation $(\frac{E}{c})^2 = \mathbf{p}^2 + (mc)^2$ could be interpreted in terms of a partial differential equation

$$\frac{1}{c^2}\left(\frac{\partial}{\partial t}\right)^2 \phi(\mathbf{x}.t) = (-\hbar^2 \nabla^2 + (mc)^2)\phi(\mathbf{x}, t), \qquad (3.2)$$

where $\phi(\mathbf{x}, t)$ is the wave function

$$\phi(\mathbf{x}, t) = \int \frac{d^3 p}{2E} e^{\frac{i}{\hbar}(\mathbf{x}\cdot\mathbf{p} - \frac{E}{c}t)} \qquad (3.3)$$

for the de Broglie waves. The derivatives in (3.2) reproduce the mass-energy relation, so one may think of of this wave function as carrying essential information about the particle.

This equation, now called the Klein-Gordon equation [Klein (1926), Gordon (1926)] was studied by Erwin Schrödinger [Born (1920)] and found to be not so useful at the time; he therefore made a non-relativistic approximation as follows. The formula

$$\left(\frac{E}{c}\right) = \sqrt{\mathbf{p}^2 + (mc)^2} \qquad (3.4)$$

or

$$E = c\sqrt{\mathbf{p}^2 + m^2 c^2}$$

$$= mc^2 \sqrt{1 + \frac{\mathbf{p}^2}{m^2 c^2}}$$

$$= mc^2 + \frac{\mathbf{p}^2}{2m} + O\left(\frac{1}{c^2}\right) \qquad (3.5)$$

provides a model for a class of problems in non-relativistic quantum mechanics.

The corresponding wave equation was then taken to be

$$i\hbar\frac{\partial}{\partial t}\psi_t(\mathbf{x}) = \frac{\mathbf{p}^2}{2m}\psi_t(\mathbf{x}), \tag{3.6}$$

where $\mathbf{p}^2$ is represented, as above, by $-\hbar^2\nabla^2$. This result is called the *Schrödinger Equation*. The left hand side corresponds to the Hamiltonian for a free non-relativistic particle. With, as in (2.45), an additional term $V(\mathbf{r})$. This equation provides a basis for non-relativistic quantum dynamics. In the limit $\hbar \to 0$, the theory would become the standard non-relativistic classical mechanics, with $i\hbar\frac{\partial}{\partial t}\psi_t(\mathbf{x}) \to E\psi_t(\mathbf{x})$. The form of the quantum Hamiltonian is then suggested by a *correspondence principle*, writing the theory in a form that would smoothly go over in this limit to the known structure of classical mechanics.

We now argue that the solution must be bounded. The integration of the radial part after separation of variables *e.g.* [Merzbacher (1970)] for the analog of the problem with Coulomb potential, for $V(r) = -\frac{e^2}{r}$, considering E, the energy, as suggested by de Broglie (1924), as a parameter as obtained from (as mentioned above)

$$i\hbar\frac{\partial}{\partial t}\psi_t(\mathbf{x}) = E\psi_t(\mathbf{x}), \tag{3.7}$$

leads to the conclusion *e.g.* [Schwinger (1954)] that the solution for radial part runs to infinity except for choices of E that coincide with the measured values of energy of the electron in the hydrogen atom. This result shows that the Schrödinger theory explains the assertions of Bohr, but also shows that the wave function must be bounded (we will show later that the integral of the *absolute square* of the function must be bounded).

The fundamental underlying reason for this quantum mechanical structure lies in the seminal thought experiment of Heisenberg. He studied the problem of finding an electron by projecting a light pulse toward the particle and examining the reflected pulse, which should contain information about the electron's position. However, during the collision process, the electron's position is perturbed with variation satisfying

$$\Delta k \Delta x \geq 1, \tag{3.8}$$

for $k = \frac{2\pi}{\lambda}$, where λ is the wavelength of the light, or, for $p = \hbar k$ [Planck (1878)],[1]

$$\Delta p \Delta x \geq \hbar. \qquad (3.9)$$

This fundamental thought experiment shows that there are pairs of classically defined quantities which are *incompatible*, *i.e.*, cannot be simultaneously measured in the same apparatus. Therefore, the Newtonian-Galilean notion of a particle following a well-defined orbit, with precisely known position and momentum at each point, no longer has objective meaning on a microscopic scale; one must reconsider from the beginning the basis of the description of dynamical systems.

At the time as these developments, several physicists, including Schrödinger, Bohr, Heisenberg, and Rutherford were visiting at Göttingen, as well as mathematicians such as Hilbert, Courant, Feller and others. In particular, Hilbert was searching for a mathematical structure for sets of functions, say, $f(x)$, which would have the properties of Euclidean geometry, for example, that the sum of the lengths of two sides of a right triangle would be greater or equal to the hypotenuse, or that the size of $f + g$ would be less than or equal to the size of f plus the size of g. The result was what is called a *Hilbert space*, with the properties that each f in the space has a finite size (norm) called $\|f\|$, and there is a *scalar product* denoted by (f, g), complex valued, for which $(f, g)* = (g, f)$ and $(f, f)^{\frac{1}{2}} = \|f\|^2$. It satisfies the Schwartz inequality (as in Euclidean geometry), $(f, g) \leq \|f\|\|g\|$. The functions in this set, called a *space*, are to have the property of completeness and closure, that is, any complex linear combination of such functions is in the set, and the limit of a sequence f_n has the property that $\|f_n - f_m\|$ is arbitrarily small for n and m arbitrarily large. Remarkably, it turned out that this structure was what is needed by the physicists to describe the new theory of quantum mechanics that was emerging.

The example that we gave of the solution of the Schrödinger equation which yielded special discrete values of the energy was a

---

[1]We shall see in a rigorous quantum theory that the right hand side should have a factor of one half.

good illustration. In fact, under certain conditions, the Hamiltonian for a physical system may have a discrete set of values, called *eigenvalues* (or, "proper values"), in addition to a continuum. We remark that there may be several functions which yield the same value for the energy; a phenomenon called degeneracy.

In general, one may define *subspaces*, sets of functions that have the same properties as the full space, and define operators which select from the complete set of functions all those of a particular subspace, for example, a degenerate set. These operators are called *projection operators.*[von Neumann (1971)].

Before concluding our mathematical introduction, we wish to state an additional result of fundamental importance, the Riesz theorem [Riesz and Nagy (2014)]. This theorem states that for any complex valued bounded linear function $L(f)$ of f, *for all f*, for which

$$|L(f)| \le K\|f\|, \tag{3.10}$$

for K some finite real number, there is a vector g such that

$$L(f) = (g, f). \tag{3.11}$$

In particular, for the operator H bounded,

$$(g, Hf) \le \|g\|\|Hf\| \le \|g\|\|\|H\|\|\|f\|, \tag{3.12}$$

which satisfies (3.10) for H bounded (here, $\|H\|$ is the maximum of (f, Hf) for all f. the bound of H). According to the Riesz theorem, then, there is a vector h such that

$$(g, Hf) = (h, f). \tag{3.13}$$

The vector h must be linear in g, so we may *define*

$$h \equiv H^\dagger g, \tag{3.14}$$

where $H^\dagger$ is called the *adjoint* (or Hermitian conjugate) of H. If $H^\dagger = H$, the operator H is called self-adjoint.[2] The operators representing physical observables are self-adjoint, guaranteeing real values upon measurement.

---

[2]In case the operator is not bounded, we may take g such that the unbounded values of Hf are orthogonal to it, thus providing a non-trivial domain for $H^\dagger$, still leaving our functional defined for all f (discussion of [Naimark (1964)]).

As discussed above, the thought experiment of Heisenberg played an essential role in the development of the quantum theory. The existence of incompatible observables, in the formulation of Birkhoff and non Neumann (1955), defined a *state* function $w(M)$ on the subspaces of the Hilbert space, carrying the likelihood of the system to be represented in each subspace. Incompatability is reflected in the fact that $w(M_1 \cup M_2) \neq w(M_1) + w(M_2)$ ($\cup$ is the union) unless M_1 is orthogonal to M_2, and $w(M_1 \cap (M_2 \cup M_3)) \neq w(M_1 \cap M_2) + w(M_1 \cap M_3)$, where $\cap$ is the intersection, unless M_2 and M_3 are orthogonal. The resulting structure is called a *lattice*.

Piron (1976), on the other hand, did not start with a Hilbert space, but defined a *measure* on the set of physical properties of a system, yes-no questions $a, b, c...$, called *propositions* that could characterize the physical properties of the system. The notion of incompatibility then arises in the property that $w(a \cup b) \neq w(a) + w(b)$ unless a and b are mutually exclusive, and $w(a \cap (b \cup c)) \neq w(a \cap b) + w(a \cap c)$ unless b and c are mutually exclusive. Piron showed that the corresponding structure can be *imbedded* in a Hilbert space, as assumed initially by Birkhoff and von Neumann. Piron's approach provides a framework for the axiomatic foundation of the quantum theory.

In the following, we provide, in some detail, another constructive approach, introduced by J.S. Schwinger (1969), which gives a deep insight into the content of the quantum theory.

Suppose that we have a beam of particles each of which has properties $A, B, C, \ldots$, which may have values $a', b', c' \ldots$. We may set up a *filter* which allows all particles with the value of a' for A to pass, independently of the values of $B, C, D, \ldots$. This filtering corresponds to a *measurement*, called $M(a')$. If we store the outcome, and add to this the outcome of the measurements searching for systems with value of A equal to $a'', a''', \ldots$ through all possible values of A (we assume a discrete set initially to develop the main ideas), we will surely reconstitute the original beam. We may represent this as the *unity* measurement, I, that passes all particles, *i.e.*,

$$\Sigma_{a'} M(a') = I. \tag{3.15}$$

An essential idea in this construction is that if we measure the outgoing beam, immediately after the first measurement for a' with a filter admitting particles with value b' of B; if B and A are not compatible, the final beam passing will have b' for B, but generally no longer a' for A. Filtering for a'' in the last step, we would have $M(a'')M(b')M(a')$, which may not be zero; the measurement of B may disturb the value found for A.

To complete the algebra, we note that $M(b')M(a')$ must be proportional to $M(b', a')$, a measurement called a *Pauli measurement of the second kind*, that selects systems with A in a' and leaves them with B in b'. We now define a coefficient, usually taken to be complex valued, called $\langle b'|a'\rangle$ [Dirac (1947)] so that

$$M(b')M(a') = \langle b'|a'\rangle M(b', a'), \qquad (3.16)$$

and the time reversed measurement

$$(M(b')M(a'))^\dagger = M(a')M(b') = \langle a'|b'\rangle M(a', b'). \qquad (3.17)$$

Furthermore, from (3.16),

$$M(a'')M(b')M(a') = \langle a''|b'\rangle\langle b'|a'\rangle M(a'', a'), \qquad (3.18)$$

a measurement causing a transition from a' to a''. If we sum over all possibilities b', as in (3.15), we see that

$$\Sigma_{b'}\langle a''|b'\rangle\langle b'|a'\rangle = \delta(a'', a'). \qquad (3.19)$$

We now define the *trace* operation as

$$Tr\, M(b', a') = \langle a'|b'\rangle, \qquad (3.20)$$

and the conjugate operation

$$Tr\, M(b', a')^* = Tr\, M(a', b'), \qquad (3.21)$$

It therefore follows that

$$\langle b'|a'\rangle^* = \langle a'|b'\rangle, \qquad (3.22)$$

so that (3.10) corresponds to *unitarity*.

It is generally assumed that the numbers $\{\langle a'|b'\rangle\}$ are complex, but the star division algebras such as quaternion or octonion algebras, or general (star) Clifford algebras, have been investigated as well [Goldstine and Horwitz (1963)]. Here we assume, as in the usual quantum theory, that these numbers are complex.

We then find that

$$\Sigma_{b'}|\langle a'|b'\rangle|^2 = 1. \tag{3.23}$$

That the sum of a set of positive numbers associated to physical properties adds to unity is strongly suggestive of a probability measure, in this case, the sum of probabilities over all values of b' of B for which A would be found with value a'.

Let us represent the *observable* A by

$$A = \Sigma_{a'}a'M(a'). \tag{3.24}$$

This definition has a simple operational consequence, that

$$AM(a') = a'M(a'). \tag{3.25}$$

Moreover, if we compute the trace with a measurement testing for B with value b', we obtain

$$Tr\,(M(b')A) = \Sigma_{a'}a'Tr(M(b')M(a')) = \Sigma_{a'}a'|\langle a'|b'\rangle|^2, \tag{3.26}$$

With this, we can define the *expectation value* of A given that B has the value b', *i.e.*,

$$\langle A\rangle_{b'} = Tr(M(b')A). \tag{3.27}$$

We may then think of the situation for which B has the value b' as the *state* of the system, the coefficients $|\langle a'|b'\rangle|^2$ correspond to the probabilities that, given B with value b', A has the value a' (obviously symmetrical).

The calculus discussed here was called by J. Schwinger, its originator, the *Measurement Algebra* (1969). The measurement symbols may be represented in terms of the bilinear form, bringing us closer to the usual form of the quantum theory,

$$M(a',b') = \psi(a')\psi(b')^\dagger \tag{3.28}$$

with the rule

$$\psi(b')^\dagger \psi(a') = \langle b'|a' \rangle. \tag{3.29}$$

Dirac (1947) has used for the objects $\psi(a')$ "kets" $|a'\rangle$ and their conjugates $\langle b'|$ "bras", from which (3.29) becomes an evident mnemonic.

Going to the continuum limit for the set $\{a'\}$, calling it, say, x, the quantity $\langle x|b' \rangle \equiv \psi_{b'}(x)$ is called the *wave function*, representing the state of the system, to be identified with an element of the Hilbert space, as described above. From the completeness relation (3.13), with a measure dx, say, on coordinate space,

$$\int dx \langle a'|x \rangle \langle x|a' \rangle = \int dx \psi_{a'}(x)^* \psi_{a'}(x) = \int |\psi_{a'}(x)|^2 = \|\psi_{a'}\|. \tag{3.30}$$

For a state with norm unity,

$$\int dx |\psi_{a'}(x)|^2 = 1. \tag{3.31}$$

We therefore see that $|\psi_{a'}(x)|^2$ can be understood as the *probability density* on the measure dx to find the particle in this interval. This is called the *Born rule* [Born (1920)].

Schwinger's formulation therefore provides a basis for the quantum theory which carries a clear interpretation of its conceptual content.

In the next section, we discuss the dynamical *evolution* of the wave function and questions of locality which arise in a relativistic framework.

3.2 Evolution of the Wave Function and Locality

In this section, we discuss some of the difficulties that arise from attempts to use the Klein-Gordon construction as a relativistic quantum theory, in particular the problems of locality and non-positive probability density. In the next section, we explain how the Stueckelberg off-shell provides a fully relativistically covariant quantum theory which is local and has positive probability. In succeeding chapters we develop some important applications.

As we have discussed, in connection with (3.5), for a system with classical Hamiltonian

$$H = \frac{\mathbf{p}^2}{2m},\tag{3.32}$$

one may write the (Schrödinger) equation

$$i\hbar\frac{\partial}{\partial t}\psi_t(\mathbf{x}) = \frac{\mathbf{p}^2}{2m}\psi_t(\mathbf{x}),\tag{3.33}$$

where $\mathbf{p}^2$ is represented as $-\hbar^2\frac{\partial^2}{\partial\mathbf{x}^2}$. The Fourier transform of $\psi_t(\mathbf{x})$ is

$$\psi_t(\mathbf{p}) = \int d^3\mathbf{x}\, e^{\frac{i}{\hbar}\mathbf{p}\cdot\mathbf{x}}\psi_t(\mathbf{x});\tag{3.34}$$

as we have seen, $|\psi_t(\mathbf{x})|^2$ is the probability density to find the particle in a small volume $d^3\mathbf{x}$. By the Parseval theorem, it then follows that $|\psi_t(\mathbf{p})|^2$ is the probability density to find the particle in a small volume of momentum space $d^3\mathbf{p}$. This property, for which the wave function describes the place that a particle can be found (both in coordinates and momenta) is called *locality*.

According to (3.31), one can write

$$\psi_t(\mathbf{x}) = (e^{\frac{i}{\hbar}Ht}\psi_{t=0})(\mathbf{x})\tag{3.35}$$

(with $t = 0$ as initial value), describing, in a probabilistic sense, the motion of the particle, in place of the classical Newtonian trajectory.

Since the Hamiltonian operator is Hermitian, it follows that

$$\|\psi_t\|^2 = \|\psi_{t=0}\|^2,\tag{3.36}$$

so that the total probability is conserved for all t for the non-relativistic theory.

For a relativistic theory, Klein and Gordon (1926) proposed returning to Schrödinger's original idea, based on the work of de Broglie. We restate the Klein Gordon equation again here, using units for which $\hbar = c = 1$,

$$(-\partial_\mu\partial^\mu + m^2)\phi(x) = (\partial_t^2 - \nabla^2 + m^2)\phi(x) = 0,\tag{3.37}$$

where $x \equiv (\mathbf{x}, t)$. The *current* (flow of particles per unit area per unit time)

$$J_\mu(x) = \frac{i}{2}(\phi^*(x)\partial_\mu\phi(x) - \partial_\mu\phi^*(x))\phi(x)) \qquad (3.38)$$

is conserved, *i.e.*

$$\partial^\mu J_\mu(x) = 0, \qquad (3.39)$$

as is easily seen from (3.37).

We may represent $\phi(x)$ in terms of its Fourier transform from momentum space, taking into account that the mass squared $-p_\mu p^\mu = E^2 - \mathbf{p}^2$ (for $p^\mu = E, \mathbf{p}$ has a definite value m^2, with (here $E \geq 0$)

$$\phi(x) = \int d^4p\,\delta(-p_\mu p^\mu - m^2)e^{ip_\mu x^\mu}\phi(p)$$

$$= \int d^4p\,\delta(E - \sqrt{(\mathbf{p}^2 + m^2)}(E + \mathbf{p}^2 + m^2))$$

$$= \int \frac{d^3\mathbf{p}}{2E}\left[e^{i\mathbf{p}\cdot\mathbf{x}-iEt}\phi(\mathbf{p}, E) + e^{i\mathbf{p}\cdot x + iEt}\phi(\mathbf{p}, -E)\right], \qquad (3.40)$$

where $E = +\sqrt{\mathbf{p}^2 + m^2}$. Assuming we have positive energy solutions only, *i.e.* $\phi(\mathbf{p}, -E) = 0$, only the first term of (3.40) contributes. Then,

$$\int d^3x\, J_0(x) = \int \frac{d^3\mathbf{p}}{2E}|\phi(\mathbf{p}, E)|^2. \qquad (3.41)$$

With the same measure as the norm, one then defines a scalar product

$$(\phi_1, \phi_2) = \int \frac{d^3\mathbf{p}}{2E}\phi_1(\mathbf{p})^*\phi_2(\mathbf{p}). \qquad (3.42)$$

Newton and Wigner (1949) then assumed there is a particle localized at the point $\mathbf{x} = 0$ with wave function $\phi_0(\mathbf{x})$, so that for

$$\phi_1 = \phi_0 \quad \text{at } \mathbf{x} = 0$$

$$\phi_2 = \phi_0 e^{-i\mathbf{p}\cdot\mathbf{a}} \quad \text{at } \mathbf{x} = \mathbf{a}, \qquad (3.43)$$

ϕ_1 should be orthogonal to ϕ_2. The scalar product between them is

$$\int \frac{d^3\mathbf{p}}{2E} |(\phi_0\mathbf{p})|^2 e^{-i\mathbf{p}\cdot\mathbf{a}} = 0, \qquad (3.44)$$

for all $\mathbf{a} \neq 0$. Therefore, the function transformed must be a constant (up to a phase, which they argue may be taken to be unity) *i.e.*,

$$\phi_0(\mathbf{p}) = C\sqrt{2E}, \qquad (3.45)$$

so that at a point $\hat{\mathbf{x}}$,

$$\phi_{\hat{\mathbf{x}}}(\mathbf{p}) = C\sqrt{2E}\, e^{-i\mathbf{p}\cdot\hat{\mathbf{x}}}. \qquad (3.46)$$

Then,

$$\phi_{\hat{\mathbf{x}}}(\mathbf{x}) = C \int \frac{d^3\mathbf{p}}{2E} e^{-i\mathbf{p}\cdot\hat{\mathbf{x}}-\mathbf{x}} \qquad (3.47)$$

This wave function for a particle at a given point is not localized (a Bessel function); it has a width of the order of $\frac{1}{m}$. For $m \to 0$ (photon or light neutrino), it spreads to ∞.

To find the *position operator*, consider again (3.46), and compute

$$i\frac{\partial}{\partial\mathbf{p}}\phi_{\hat{\mathbf{x}}}\mathbf{p}) = C\sqrt{2E}\,\mathbf{x}e^{-i\mathbf{p}\cdot\hat{\mathbf{x}}} + iC\sqrt{2E}\frac{1}{2}\frac{\mathbf{p}}{2E^2}e^{-i\mathbf{p}\cdot\mathbf{x}}, \qquad (3.48)$$

providing the result

$$\mathbf{x}_{NW} = \left(i\frac{\partial}{\partial\mathbf{p}} - \frac{i}{2}\frac{\mathbf{p}}{E^2}\right) \qquad (3.49)$$

acting on the wave function (3.46).

Multiplying all wave functions in the Hilbert space by $\frac{1}{\sqrt{2E}}$, the operator $\mathbf{x}_{NW}$ becomes just $(i\frac{\partial}{\partial\mathbf{p}})$, in the usual local form. This is actually the transformation of Foldy and Wouthysen (1950) for scalar fields. The resulting wave functions, however, then are no longer covariant. We shall see in the next section that in the relativistic quantum theory of Stueckelberg, Horwitz and Piron (2015), the wave functions are local and covariant.

3.3   The Stueckelberg-Horwitz-Piron Quantum Theory

The conflict between covariance and locality, as well other difficulties
which arise in following the construction of a quantum theory
of the form discussed in the previous subsection can be resolved
by building a quantum theory based on the classical relativistic
dynamics discussed in Chapter 2.

The classical Poisson bracket relations, leading to the canonical
form (2.53) can be taken as the basis of covariant quantum theory,
following Dirac (1930) by assuming the *operator relations*

$$[x^\mu, p_\nu] \equiv x^\mu p_\nu - p_\nu x^\mu = i\hbar \eta^\mu{}_\nu, \tag{3.50}$$

where the symbols x^μ and p_ν correspond to self-adjoint operators
with spectra that contain the outcomes of measurements of the
observables, the spacetime coordinates and energy-momentum of
the particle at a given (universal) time τ. These operator relations
directly imply that the operator valued form of $M_{\mu\nu}$ of (2.54)
transforms these variables according to the Lorentz group, resulting
in a completely covariant quantum theory. One can then write the
so-called *Stueckelberg-Schrödinger equation equation* (for a potential
model as in (2.44))

$$i\hbar \frac{\partial \psi_\tau(x)}{\partial \tau} = \left(\frac{1}{2M} p^\mu p_\mu + V(x) \right) \psi_\tau(x), \tag{3.51}$$

where p_μ, is represented as $i\hbar \frac{\partial}{\partial x^\mu}$, and x corresponds to the operators
x and t. The spectrum of these operators contains the *measured*
values of **x** and the time t of emergence of the event in the detector.
The wave functions, covariant elements of the Hilbert space, may be
normalized according to

$$\int d^4x |\psi_\tau(x)|^2 = 1, \tag{3.52}$$

integrated over all spacetime, with corresponding scalar product

$$(\psi_{1\tau}, \psi_{2\tau}) = \int d^4x \psi_1{}^*{}_\tau(x) \psi_{2\tau}(x). \tag{3.53}$$

At each τ, the wave function (as for $\psi_t(\mathbf{x})$ in the nonrelativistic
theory) is *coherent* on the manifold, in this case $\{\mathbf{x}, t\}$. As for the

nonrelativistic theory, where the wave function is coherent in **x**, one can predict interference in the spatial double slit experiment [double slit, quantum interference, Davisson and Germer (1928)], in the relativistic theory, we can predict *interference in time* as well for an experiment involving gates in time [Horwitz (1976)], as for the laser pulses on Argon gas in the Lindner experiment (2005). We will discuss this experiment and its interpretation in the next section.

The Fourier transform (we use the same symbol for the function of x and of p, distinguishing them by their arguments)

$$\psi_\tau(p) = \frac{1}{(2\pi)^2} \int d^4x\, e^{-ix^\mu p_\mu} \psi_\tau(x), \qquad (3.54)$$

satisfies the Parseval relation

$$\int d^4x\, |\psi_{1\tau}(x)|^2 = \int d^4p\, |\psi_\tau(p)|^2, \qquad (3.55)$$

expressing the invariance of the norm $\|\psi_\tau\|$ in either representation. As we have seen in the discussion of the measurement algebra, the Fourier kernel, $\langle x|p\rangle = \frac{1}{(2\pi)^2} e^{-ix^\mu p_\mu}$ as for $\langle a'|b'\rangle$ defined in (3.16), acts as a *transformation function* from the x to the p representation.

3.4 The Lindner Experiment

In this section, we describe the Lindner experiment (2005) and its implications for the nature of the observed time. It was the first experiment showing coherence in time, crucial for the SHP theory [Horwitz (2015)].

In this experiment, laser light of about 850 nm wavelength is radiated onto a sample of Argon gas in a short pulse of one and a half wavelengths, constituting two peaks in the electric field in one direction, and one, in between, in the opposite direction. An electron may be emitted as a result of interaction with the first peak, or the third, separated by about one femtosecond in time. At the detector, one sees an interference pattern between the two possiblities corresponding to ejection at the first or third maximum in the wave, much like the double slit experiment in space [Merzbacher (1970)]. The second peak in the opposite direction, which exhibits no perceptible interference effect, was used to confirm that just a single electron was involved in the process.

The interference observed in the spacial double slit experiment is accounted for by the coherence of the wave function in space, and was one of the earliest experimental confirmations [Davisson (1927)] of the structure of the quantum theory as it emerged from its formulation in Hilbert space. In view of the development of recent technologies, it was natural for the group at the Max Planck Institute [Lindner *et al.* (2005)] to ask whether one could see interference in time. Their experiment was remarkably successful, but raised fundamental questions on the role of time in the quantum theory.

This experiment clearly shows the effect of quantum interference in time. The results are discussed in that paper in terms of a very precise solution of the time-dependent nonrelativistic Schrödinger equation. However, the nonrelativistic Schrödinger theory cannot, according to the basic principles of the quantum theory, be used to predict interference phenomena in time, and therefore the very striking results of this beautiful experiment have a fundamental importance which goes beyond the technical advances which they represent. These results imply, in fact, that the time variable t must be adjoined to the set of standard quantum variables so that the standard ket $|x,t\rangle$ for the representation of the quantum state (in Dirac's terminology [Dirac (1930)]) can be constructed [Horwitz (2006)]. It is this structure for the wave function $\psi(x,t) \equiv \langle x,t|\psi\rangle$, where x and t are the spectra of self-adjoint operators, that provides the possibility of coherence in t, and therefore, interference phenomena. If the quantum theory is to remain symplectic in form, moreover, the variable E (in addition to t) must also be adjoined.

Let us examine the reasons why the standard nonrelativistic quantum theory cannot be used to predict interference in time. For example, Ludwig (1982) has pointed out that "time" cannot be a quantum observable, since there is no observable that does not commute with t in the nonrelativistic theory. Note that the Hamiltonian of the standard theory evolves quantum states in time, but does not act as a shift operator since it commutes with t. Moreover, Dirac (1947) has argued that if t were an operator,

then the resulting t, E commutation relation would imply that the
energy of the system is unbounded below (with no gaps), from
which he concluded that the time cannot be an observable in the
nonrelativistic quantum theory.

As the axiomatic treatment of Piron (1976) (see also, [Jauch
(1968)]) shows, the Hilbert space of the quantum theory is constructed of a set of wave functions satisfying a normalization
condition based on integration over all space, *e.g.*, for a single
particle, $\int |\psi_t(x)|^2 d^3 x \leq \infty$, for each value of the parameter t. Since
t is not integrated over, ψ_t does not carry a probability distribution
for values of t. There is a distinct Hilbert space for each value of the
parameter t.

Moreover, as pointed out by Wick, Wightmann and Wigner
[Wick (1952)], a Hilbert space decomposes into incoherent sectors
if there is no observable that connects these sectors; hence, if there
were a larger Hilbert space containing a representation for t, the
absence of any observable that connects different values of t in the
standard nonrelativistic physics would induce a decomposition of
the Hilbert space into a (continuous) direct sum of superselection
sectors [Piron (1976)]. Therefore, no superposition of vectors for
different values of t would be admissible. This argument would
also would exclude the interpretation of the experiment given by
Lindner *et al.* (2005) forming the basis of the analysis carried out
by the authors involving the linear superposition of two parts of a
particle wave function arriving at the detector simultaneously, but
originating at two different times, in the framework of the standard
nonrelativistic quantum theory.

The situation for particles, in this respect, is very different from
that of electromagnetic waves, for which the second order equations
imply coherence in time as well as space (the coherence time for
light waves is a commonly measured characteristic of light sources).
It is clear from the (spatial) double slit interference of light, which
travels at a fixed velocity, that the sections of a wave front passing
through the two slits must pass at different times if they are to arrive
simultaneously at the detection plane off-center. The arrival of pieces
of a *particle* wave packet which have passed through two spatially

separated slits simultaneously on a screen off center is made possible
by the dispersion of momenta in the wave packet, permitting a range
of velocities. If the two contributions to the linear superposition on
the screen were not taken to be simultaneous at the two slits, they
would not interfere, since they would have originated on wave packets
at different values of time. If, indeed, such interference could take
place, we would have to add up the contributions passing each slit
for all times, and this would destroy the interference pattern (one can
see in the standard calculation in every textbook [*e.g.* Merzbacher
(1970)] that the two pieces of the wave packet that contribute to the
interference are taken at equal time, *i.e.*, from a single wave packet
arriving at the slits).

We now show that a simple argument based on the propagator
for the Schrödinger equation demonstrates that interference in time
could not occur in the standard Schrödinger theory in a simple and
general way.

The free propagator for wave functions in the standard
Schrödinger treatment is given by

$$\langle x|U(t)|x'\rangle = \left(\frac{m}{2\pi i t}\right)^{\frac{3}{2}} e^{i\frac{m}{2(t-t')}(x-x')^2}$$
$$= G(x - x', t - t'), \tag{3.56}$$

where x is here a three dimensional variable. The action of this
propagator, a transcription of the Schrödinger evolution $e^{-iH(t-t')}$
into coordinate representation, is

$$\psi_t(x) = \int dx' G(x - x', t - t')\psi_{t'}(x'). \tag{3.57}$$

The integration over x makes possible the description of the double
slit experiment in space (by coherently adding up contributions from
two or more locations in x at a given t to the wave arriving at a
screen over $\{x'\}$ at the time t'); there is, however, no integration
over t, and therefore no mechanism for constructing interference in
time. This result, obvious from the form of (3.57), is a reflection of
the arguments of Ludwig (1982) cited above, and is fundamental to
the standard nonrelativistic quantum theory. This result constitutes
a formal argument that no interference effect in time is predicted

by the standard nonrelativistic Schrödinger theory. Introducing two packets into the beam of an experiment at two different times t_1 and t_2 would result in the *direct sum* of the two packets at some later time, say t_3, if one propagates the first from t_1 to t_3, and the second from t_2 to t_3. This would constitute, by construction, a mixed state, for which no interference would take place, just as the construction of a beam of $n+m$ particles by adding a set of n particles with definite spin up to another set of m particles with spin down results in a mixed state. Such a mixed state is described by a diagonal density matrix with *a priori* probability $n/(n+m)$ for outcome spin up, and $m/(n+m)$ for outcome spin down. There is no measurement which would result in some spin with certainty in *any* direction [Jauch (1968)].

To understand the apparent qualitative success of the calculations performed by the Lindner group [Lindner (2005)], suppose we assume that we are dealing with ordinary functions (as distinguished from the rays in Hilbert space describing a quantum state [Wigner (1931); see also Weinberg (1995)]) described by the Schrödinger evolution, and carry out an approximate calculation for short time intervals, ignoring the fact that these contributions cannot be coherent.

If we were to assume, arbitrarily, that the waves from sources at two different times could be coherently added, the propagator above could be expanded in a power series for small variations in the final time, and one would find some semblance of an interference pattern for a few maxima before distortions would set in. For ε the time between peaks of the laser beam, $\mathcal{T}$ the time between peaks on the predicted "interference pattern" on a screen at a distance L from the emission source, m the mass of the electron, one finds a crude estimate (in agreement with the calculations of Lindner *et al.*)

$$\varepsilon\mathcal{T} \cong \pi\hbar\sqrt{\frac{m}{2}}L(E_{kin}^e)^{-\frac{3}{2}}, \qquad (3.58)$$

where E_{kin}^e is the kinetic energy of the emitted electron, and L is the distance from emitter to the detector. It is instructive to understand how this formula is obtained from the nonrelativistic Schrödinger equation,

$$\frac{i}{\hbar}\frac{\partial\psi_t(\mathbf{x})}{\partial t} = H\psi_t(\mathbf{x}), \qquad (3.59)$$

where we shall take H to be the free Hamiltonian $\mathbf{p}^2/2m$. The formal solution of this equation is

$$\psi_t(\mathbf{x}) = \left(e^{-iH(t-t')/\hbar}\psi_{t'}\right)(\mathbf{x}) \tag{3.60}$$

where $\psi_{t'}$ represents the state at time t'. The matrix element of the unitary transformation appearing in Eq. (3.60),

$$\langle \mathbf{x}|e^{\frac{-iH(t-t')}{\hbar}}|\mathbf{x}'\rangle = \frac{1}{(2\pi\hbar)^3}\int d^3p\, e^{i\mathbf{p}\cdot(\mathbf{x}-\mathbf{x}')/\hbar}e^{-i\frac{\mathbf{p}^2}{2m}(t-t')/\hbar} \tag{3.61}$$

where we have used the transformation function

$$\langle \mathbf{x}|\mathbf{p}\rangle = \frac{1}{(2\pi\hbar)^{\frac{3}{2}}}e^{i\mathbf{p}\cdot\mathbf{x}/\hbar} \tag{3.62}$$

to transform from the $\mathbf{x}$ representation to the $\mathbf{p}$ representation, for which the Hamiltonian is diagonal. The resulting integration can be easily carried out to yield (as in (3.56))

$$\langle \mathbf{x}|e^{\frac{-iH(t-t')}{\hbar}}|\mathbf{x}'\rangle = \left(\frac{m}{2\pi i\hbar(t-t')}\right)^{\frac{3}{2}}e^{i\frac{(\mathbf{x}-\mathbf{x}')^2}{2\hbar}\frac{m}{t-t'}}. \tag{3.63}$$

We first use this result to compute the standard interference pattern observed from two slits separated in space in order to be able to directly compare the computation of space interference and time interference (assuming coherence in time as well). We use this propagator, or Green's function, to propagate the waves from two slits set apart a short distance in space at the points $\mathbf{x}_1$ and $\mathbf{x}_2$. Assuming equal phase of the wave function at the two points, the resulting wave at the point $\mathbf{x}$ in the screen will be proportional to

$$e^{i\frac{(\mathbf{x}-\mathbf{x}_1)^2}{2\hbar}\frac{m}{t-t'}} + e^{i\frac{(\mathbf{x}-\mathbf{x}_2)^2}{2\hbar}\frac{m}{t-t'}}. \tag{3.64}$$

Factoring out the overall phase given by the first term, one is left with a formula for the intensity on the screen proportional to

$$\left|1 + e^{i\left(\frac{(\mathbf{x}-\mathbf{x}_2)^2}{2\hbar} - \frac{(\mathbf{x}-\mathbf{x}_1)^2}{2\hbar}\right)\frac{m}{t-t'}}\right|^2 \tag{3.65}$$

Writing the difference between two squares in the exponent as the sum and difference, we see that the $\mathbf{x}$ dependence cancels from the difference, and in the sum factor, it is doubled, *i.e.*, the phase is given by

$$e^{i\frac{(2\mathbf{x}-(\mathbf{x}_1+\mathbf{x}_2))(\mathbf{x}_1-\mathbf{x}_2)}{2\hbar}\frac{m}{t-t'}}. \tag{3.66}$$

The cross term in the absolute square in Eq. (3.65) is twice the cosine of the angle appearing in the exponent, and we see that as a function of $\mathbf{x}$ on the image plane, there is an oscillation frequency of

$$\frac{2(\mathbf{x}_1-\mathbf{x}_2)}{2\hbar}\frac{m}{t-t'}, \tag{3.67}$$

where the quantity $(t-t')$ can be estimated by the distance to the screen divided by the average velocity of the wave packets.

In this calculation, it is clearly seen that the formation of the interference pattern is a simple consequence of the factorization of the phase factors in the numerator of the exponent.

The expected pattern to be obtained from Eq. (3.63) in the superposition of sources at two times, following the assumption of coherence, is not quite so straightforward to obtain. To do this, let us examine the corresponding linear superposition of waves, assuming coherence, from two gates in time at t_1 and t_2. Neglecting the modification of the coefficients of the wave functions that pass at these two times, the superposed wave is approximately proportional to

$$e^{-i\frac{(\mathbf{x}-\mathbf{x}')^2 m}{2\hbar(t_1-t)}} - e^{-i\frac{(\mathbf{x}-\mathbf{x}')^2 m}{2\hbar(t_2-t)}}$$

so that, factoring out the first term, up to a phase we obtain

$$1 + e^{i\frac{(\mathbf{x}-\mathbf{x}')^2 m}{2\hbar}\left(\frac{1}{(t_1-t)}-\frac{1}{(t_2-t)}\right)}. \tag{3.69}$$

This function is clearly not simply related to what might be expected to be an interference phenomenon, *i.e.*, as in (6.11) However, let us expand this result to first order in both of the small quantities $\varepsilon = t_2 - t_1$ and in the deviation $T - T_0$ from the classical flight pattern, in which both of the signals would arrive at the screen at time T_0. Let us first define

$$
\begin{aligned}
\Delta(t') &= \frac{1}{t_1 - t'} - \frac{1}{t_2 - t'} \\
&= \frac{1}{t_1 - t'} - \frac{1}{t_1 + \varepsilon - t'} \\
&\cong \frac{\varepsilon}{(t_1 - t')^2}.
\end{aligned}
\tag{3.70}
$$

Now, consider the arrival times $t' = T, T_0$; the difference in the resulting phases is given by

$$
\Delta(T) - \Delta(T_0) \cong -\frac{2\mathcal{T}}{(t_1 - T_0)^3},
\tag{3.71}
$$

where $\mathcal{T} = T - T_0$. For a 2π shift in overall phase, we must therefore have

$$
(\mathbf{x} - \mathbf{x}')^2 \frac{m}{2\hbar} \left(\frac{2\mathcal{T}\varepsilon}{(t_1 - T_0)^3} \right) = 2\pi,
$$

or,

$$
\varepsilon\mathcal{T} \cong \frac{2\pi\hbar}{m} \frac{(t_1 - T_0)^3)}{(\mathbf{x} - \mathbf{x}')^2}.
\tag{3.72}
$$

We now estimate $T_0 - t_1 \sim L/v = \dfrac{L}{\sqrt{\frac{2E^e_{kin}}{m}}}$, for $L = |\mathbf{x} - \mathbf{x}'|$, obtaining the result Eq. (3.58).

3.5 The Landau-Peierls and Newton Wigner Problems

Having defined the manifestly covariant quantum theory, we are now in a position to re-examine the Newton-Wigner problem [Newton (1949)]. From the viewpoint of this theory, we shall be able to understand the way the problem arises in the framework of theories which use equations of the type of those of Klein-Gordon and Dirac that impose a strict mass shell requirement.

We will show that the $\mathbf{x}$ operator in the Stueckelberg theory, corrected to extrapolate the occurrence of an event at some point in spacetime back to $t = 0$, as sought by Wigner and Newton, is exactly the Newton-Wigner position operator on each mass value (in the sense of a direct sum) under the integral defining the expectation value [Horwitz (1973)].

Consider the expectation value of $\mathbf{x}$:

$$\langle x \rangle = \int d^4 p \psi^*(\mathbf{p}, E) i \frac{\partial}{\partial \mathbf{p}} \psi(\mathbf{p}, E) \tag{3.73}$$

We now change variables, considering only $E \geq 0$, using the relation

$$E = \sqrt{\mathbf{p}^2 + m^2} \tag{3.74}$$

for m a new variable. Then,

$$dE = \frac{dm^2}{2E}, \tag{3.75}$$

where now E stands for the relation (3.74). Furthermore, if we want to think of the derivative in (3.73) as a straightforward derivative (it only acted on the first three arguments in ψ before the change of variables), we have to correct for its action on the fourth argument E, *i.e.*, we must now write

$$i \frac{\partial}{\partial \mathbf{p}} \rightarrow i \frac{\partial}{\partial \mathbf{p}} - i \frac{\partial E}{\partial \mathbf{p}} \frac{\partial}{\partial E}$$

$$= i \frac{\partial}{\partial \mathbf{p}} - i \frac{\mathbf{p}}{E} \frac{\partial}{\partial E}$$

when acting on $\psi(\mathbf{p}, E = \sqrt{p^2 + m^2})$.

We recognize that this extra term looks like velocity times time, the operator $i\partial/\partial E$. This corresponds to the displacement to get back to where a (virtual) world line would be at $t = 0$, if one imagines the semiclassical picture of a world line running through the point $(\mathbf{x}, t)$. This semiclassical interpretation of these operators, where the real information is encoded in the wave function, appears to be consistent. This extra term, however, in the quantum theory, should be symmetrized, so let us define the relativistic operator form of the Newton–Wigner operator in the context of the Stueckelberg theory as

$$x_{NW} = i \frac{\partial}{\partial \mathbf{p}} - \frac{1}{2} \{\mathbf{v}, t\}, \tag{3.76}$$

where $\mathbf{v} = \mathbf{p}/E$ and $t = i\partial/\partial E$. Then one must use the fact that when $\partial/\partial E$ acts on $\mathbf{p}/E$, it differentiates both this factor and the wave function that implicitly follows it, *i.e.*,

$$\frac{1}{2}\{\mathbf{v}, t\} = i\frac{\mathbf{p}}{E}\frac{\partial}{\partial E} - i\frac{\mathbf{p}}{2E^2}.$$

This is just the extra piece that came from the change of variables, plus a new term, which we saw is part of the Newton–Wigner operator. Thus, our operator (3.76), put into expectation value, can be seen as the expectation value of

$$\mathbf{x} \to \mathbf{x} - i\frac{\mathbf{p}}{2E^2},$$

as required by Newton and Wigner, but under the integral over all mass shells.

Therefore, the operator (3.76) may be represented as the Newton–Wigner operator under the integration over masses of an expectation value at each value of m. The semiclassically expected value of the position of a particle as it passes $t = 0$ corresponds in this way to the Newton–Wigner operator.

We can understand from the point of view of the relativistic theory that position and mass, as the operator $\mathbf{x}$ and $m = \sqrt{E^2 - \mathbf{p}^2}$, are not compatible. The Klein Gordon theory does not consider the mass to be an operator; it is just a given number, corresponding to a point on the continuous spectrum of m. The Stueckelberg theory is completely local, consistent with our construction (3.76), and the interference phenomena we describe with the associated wave functions should predict the actual outcome of experiments. Such interference effects, predicted by Horwitz and Rabin (1976), have indeed been observed.

In 1931, Landau and Peierls [Landau(1931)] deduced a relation between dispersion in momentum and time of the form (we restore $\hbar$ and c in several formulas of this section to make the units clear)

$$\Delta p \Delta t \geq \hbar/c \tag{3.77}$$

concerning the time interval Δt during which the momentum of a particle is measured and the momentum dispersion of the state.

According to Landau and Peierls, for any given dispersion of momentum in the state, there is a minimum interval of time necessary for measuring the outcomes predicted by knowledge of the state consistent with the relativistic bound on the velocities.

Landau and Peierls begin with the estimates of first order perturbation theory for the "almost conservation of energy", *i.e.*,

$$|E - E'| \sim \hbar/\Delta t; \tag{3.78}$$

where, in perturbation theory, one argues that in sufficient time Δt, the initial energy E and the final energy E' after the transition are close. This relation corresponds to the well known estimate for the nonrelativistic energy time uncertainty relation.

Landau and Peierls, however, use this result, not a rigorous property of the wave functions of a particular state, to argue that if there is a dispersion in energy in the incoming state, and a dispersion in the outgoing state, the two sets of values must be restricted by this relation, for which the central values essentially cancel. Thus, one obtains

$$|\Delta E - \Delta E'| \sim \hbar/\Delta t. \tag{3.79}$$

They then use the relation (valid for both nonrelativistic and relativistic kinematics)

$$\Delta E = \frac{dE}{dP}\Delta P = v\Delta P; \tag{3.80}$$

using absolute conservation of momentum to assert that

$$\Delta P = \Delta P',$$

they then obtain

$$|(v - v')|\Delta P \sim \hbar/\Delta t. \tag{3.81}$$

This result implies a change in velocity from incoming to outgoing states. For a given ΔP, the smaller the time interval of measurement, the larger this velocity change must be. It is however, bounded by the velocity of light c, and one therefore obtains the relation (2.39).

Aharonov and Albert (Aharonov (1981)) have understood this result in terms of causality. They argue that if a measurement is made in a short time Δt which restricts the particle to a range of momenta ΔP, the wave function must extend to $\Delta x \sim (\hbar/2\Delta P)$. The Landau–Peierls result then assures that $\Delta x \leq (c/2)\Delta t$. From the point of view of Aharonov and Albert, involving causality, as well as the use of a relativistic bound by Landau and Peierls, it is clear that the relation (2.39) should be associated with relativity.

Following the method used by Landau and Peierls for the relativistic Stueckelberg-Schrödinger equation, it would follow in the same way from first order perturbation theory that

$$|K - K'| \sim \hbar/\Delta\tau \tag{3.82}$$

Since $p^\mu p_\mu = -(\frac{E}{c})^2 - \mathbf{p}^2 = m^2 c^2$, where m is the mass of the particle measured in the laboratory. The initial and final free Hamiltonians have the form

$$K = \frac{p^\mu p_\mu}{2Mc^2} = -\frac{m^2 c^2}{2Mc^2} = -\frac{m^2}{2M}$$

and therefore the relation (3.82) becomes, for small Δm,

$$\left|\frac{m^2 - m'^2}{2M}\right| \sim \frac{\hbar}{\Delta\tau}$$

$$= \frac{|(m - m')|(m + m')}{2M} \cong |\Delta m|,$$

for m close to its "mass shell" value M. We therefore find the relation [Burakovsky (1996)]

$$\Delta m \Delta\tau \cong \hbar, \tag{3.83}$$

a mass-τ uncertainty relation. This result provides a justification for the generally assumed relation that the width of the mass dispersions of elementary particles as seen in decay modes is associated with the lifetime of the particle in its proper frame. If the particle is off shell due to additional interactions during the decay process, there would clearly be corrections.

As we have noted, such estimates are not rigorous, but carry the same semiquantitative arguments used by Landau and Peierls, based on first order perturbation theory.

The $\Delta E \Delta t$ uncertainty relation in the SHP relativistic theory, on the other hand, follows rigorously from the commutation relation

$$[E, t] = i\hbar. \tag{3.84}$$

It is a general theorem in quantum mechanics that the dispersions of two self adjoint operators A and B in a given quantum state, defined by

$$\Delta A = \sqrt{\langle (A - \langle A \rangle)^2 \rangle}$$

and

$$\Delta B = \sqrt{\langle (B - \langle B \rangle)^2 \rangle}$$

are related by

$$\Delta A \Delta B \geq \frac{\hbar}{2} \left| \langle [A, B] \rangle \right|.$$

It therefore follows from (2.46) that, as a rigorous property of the wave function representing the state of the system,

$$\Delta E \Delta t \geq \hbar/2. \tag{3.85}$$

In a similar way, it is possible to show that there is a simple and rigorous derivation of (3.77) in the framework of the manifestly covariant quantum theory we are working with here.

We have seen that the results of Newton and Wigner can be obtained in a straightforward way by defining an effective Newton–Wigner operator as in (3.76), with the semiclassical meaning of an extrapolation of the event position back to the value it would have at $t = 0$, interpreting the virtual velocity field contained in the wave function as associated (in expectation value) with an actual distribution that could be thought of as a collection of possible world lines. In the same way, we can construct an *effective time operator* by extrapolating the time of observation of an event back to the $x = 0$ axis, which one might think of as the location of a Geiger counter triggered by the passage of a world line through its position at $x = 0$. We therefore define a Landau–Peierls time operator

[Arshansky (1985)] as

$$t_{LP} = t - \frac{1}{2}\left\{\mathbf{x}; \frac{\mathbf{p}E}{p^2}\right\} \qquad (3.86)$$

where $\frac{\mathbf{p}E}{p^2}$ is an *inverse velocity operator*, providing a shift in time for a virtual worldline (the semicolon implies both dot product as well as anticommutator). It then follows that

$$[t_{LP}, p] = -[\mathbf{x}, p] \cdot \frac{\mathbf{p}E}{p^2}.$$

But $(p \equiv \sqrt{\mathbf{p}^2})$

$$[x_i, p] = i\hbar \frac{p_i}{p},$$

so that

$$[t_{LP}, p] = -i\hbar \frac{E}{p}. \qquad (3.87)$$

It therefore follows that

$$\Delta t_{LP} \Delta p \geq \frac{1}{2}\hbar \langle E/p \rangle. \qquad (3.88)$$

The quantity E/p is the magnitude of the inverse velocity operator; if the virtual velocity p/E is bounded within the wave packet by the velocity of light c, we obtain the Landau–Peierls bound (2.39) as a rigorous property of the wave function describing the state of the system. There is, in principle, however, no bound on the occurrence of components of the wave function with values of p/E greater than one. On the other hand, application of the Ehrenfest theorem [Ehrenfest (1927)], when it is valid, would rule out this possibility for the same causal reasons given by Landau and Peierls. The Ehrenfest theorem for the relativistic theory has the same structure as in the nonrelativistic theory, resulting in the classical Hamilton equations for the motion of the peak of the wave packet in spacetime. We review the argument in the following.

Consider a wave packet of the form (for free evolution)

$$\psi_\tau(x) = \frac{1}{(2\pi)^2}\int e^{ip^\mu x_\mu - i\frac{p^\mu p_\mu}{2M}\tau}\chi(p), \qquad (3.89)$$

where $\chi(p)$, the momentum representation of the state, is a fairly sharp distribution in p^μ. The function $\chi(p)$ is modulus square normalized to one over integration on all four momenta if $\psi(x)$ is modulus square normalized to one over spacetime. For large τ, if one may assume that the values of x^μ also become large, the stationary phase values

$$x^\mu \sim \frac{p^\mu}{M}\tau \qquad (3.90)$$

make the primary contribution, as in the nonrelativistic argument. The value of p^μ under the integral that contributes corresponds to the sharp peak value of the momentum space wave function, and the corresponding peak in the x^μ wave function describes the motion of a classical event, as described above. In this case, a strong presence of spacelike momenta in the wavepacket could result in the evolution of the wordline in a spacelike direction, *i.e.*, with $\frac{p}{E}$ exceeding light velocity. We could therefore, on the same causal grounds as Landau and Peierls, argue that $\langle E/p \rangle$ must be greater than $1/c$, rule out such a configuration, and arrive at the Landau–Peierls relation from (3.88).

However, as Zaslavsky (1985) has pointed out in the context of the nonrelativistic theory, the conditions for the validity of the Ehrenfest theorem degrade (in this case as a function of τ) due to the spreading of the wave packet as well as the effect of interactions on the structure of $\chi(p)$. Zaslavsky (1985) called the time for validity of the Ehrenfest theorem the "Ehrenfest time", and argued that for quantum systems for which the classical Hamiltonian induces chaotic behavior the Ehrenfest time is less. Therefore, dynamical effects may occur in the relativistic theory which could result in deviations from the Landau–Peierls bound.

In the classical construction of Stueckelberg (1941) in Fig. 1, the worldline of the particle passes through a region which is spacelike. In this region, the corresponding Landau–Peierls bound would be violated, with the contrary inequality

$$\Delta p \Delta t < \hbar/c, \qquad (3.91)$$

implying that the wave function could be arbitrarily narrow in the t-direction for a given p distribution. Thus, this diagram could be described by a quantum wave packet which has normal Ehrenfest form for the incoming and outgoing lines, but may have a vertex which is very sharp in t over a small but finite distance. The spacetime diagrams discussed by Feynman (1949) may be thought of as an idealization of this limit.

The relation (3.86) was constructed from a semiclassical interpretation of the quantum observables, a procedure that was justified in our study of the Newton–Wigner problem. In that case, we began with the straightforward computation of the expectation value of the $\mathbf{x}$ operator, which has the same representation as in the nonrelativistic quantum mechanics. However, there is no corresponding analog in nonrelativistic quantum mechanics for a *time operator*; in the nonrelativistic quantum theory, t is a parameter of evolution, and its expectation value is a trivial identity [Ludwig (1982), Dirac (1930)]. We can, however, construct an argument analogous to that used for the Newton–Wigner problem within the framework of the relativistic theory, and show in the same way that the Landau–Peierls time operator (3.86) emerges from the mass-shell restriction of the expectation value of the relativistic time operator. To see this, consider the expectation value

$$\langle t \rangle = \int d^4p \, \psi^*(\mathbf{p}, E) \left(-i\frac{\partial}{\partial E} \right) \psi(\mathbf{p}, E), \qquad (3.92)$$

where we shall consider, for each value of m the magnitude of the momentum to be a function of E. Let us change the variables p^μ to the form (Ω, p, E), where Ω corresponds to the angular coordinate variables of $\mathbf{p}$, and define

$$p = \sqrt{E^2 - m^2}. \qquad (3.93)$$

Then,

$$d^4p = p^2 d\Omega dp dE = -\frac{1}{2} p d\Omega dE dm^2. \qquad (3.94)$$

We may then write

$$\langle t \rangle = -\frac{1}{2} \int p \, d\Omega \, dE \, dm^2 \psi^* \left(\sqrt{E^2 - m^2}, \Omega, E \right)$$

$$\left[-i\frac{\partial}{\partial E} \psi \left(\sqrt{E^2 - m^2}, \Omega, E \right) + i\frac{E}{p}\frac{\partial}{\partial p} \psi(p, \Omega, E) \big|_{p=\sqrt{E^2 - m^2}} \right],$$

$$(3.95)$$

where the last term (containing the factor $(\partial p / \partial E = E/p)$ compensates for the fact that after the change of variables, $i\partial/\partial E$ acts on p as well as the last argument.

We now note that the Landau–Peierls operator (3.86) can be written as

$$t_{LP} = t - \frac{1}{2} \left[i\frac{\partial}{\partial \mathbf{p}} \cdot \frac{\mathbf{p}E}{p^2} + i\frac{\mathbf{p}E}{p^2}\frac{\partial}{\partial \mathbf{p}} \right]$$

$$= -i\frac{\partial}{\partial E} - \frac{i}{2}\frac{E}{p^2} - i\frac{\mathbf{p}E}{p^2} \cdot \frac{\partial}{\partial \mathbf{p}}, \qquad (3.96)$$

where we have used the fact that (most simply, carrying this out component by component)

$$\frac{\partial}{\partial \mathbf{p}} \cdot \frac{\mathbf{p}E}{p^2} = \frac{E}{p^2}.$$

If we take the expectation value of t_{LP} in place of t, one sees that the last term cancels with the last term in (3.96), resulting in

$$\langle t_{LP} \rangle = -\frac{1}{2} \int p \, d\Omega \, dE \, dm^2 \psi^* \left(\sqrt{E^2 - m^2}, \Omega, E \right)$$

$$\times \left[-i\frac{\partial}{\partial E} - \frac{i}{2}\frac{E}{p^2} \right] \psi \left(\sqrt{E^2 - m^2}, \Omega, E \right) \qquad (3.97)$$

We now follow an argument similar to that used above for the Newton-Wigner problem to find the wave function of an event which occurs at a definite sharp *time*.

If $\psi_{t=0}(p)$ corresponds to a state for which an event is strictly localized to a point in time $t = 0$, the wave function $\psi_{t=t_0}$ must be orthogonal to it for $t_0 \neq 0$. Therefore,

$$\int d^4 p \psi^*_{t=t_0}(p) \psi_{t=0}(p) = 0 \tag{3.98}$$

for $t_0 \neq 0$. However, using the Poincaré group property $\psi_{t=t_0}(p) = e^{iEt_0}\psi_{t=0}$, we have

$$\int d^4 p e^{-iEt_0} |\psi_{t=0}(p)|^2 = 0, \tag{3.99}$$

implying that

$$\int d^3 p |\psi_{t=0}(p)|^2 = const \times (E),$$

or,

$$\int d\Omega p^2 dp |\psi_{t=0}(p)|^2 = const \times E. \tag{3.100}$$

But, as pointed out above, $p^2 dp = -(1/2)pdm^2$, so (3.100) becomes

$$-\frac{1}{2}\int pdm^2 d\Omega |\psi_{t=0}(p)|^2 = const \times E. \tag{3.101}$$

If the mass of the particle is concentrated at some value of m we conclude that

$$\int d\Omega |\psi_{t=0}(p)|^2 = \frac{1}{p} \times const, \tag{3.102}$$

or, for a spherically symmetric wave function,

$$\psi_{t=0}(p) \propto \frac{1}{\sqrt{p}}.$$

Shifting by translation in t, we see that

$$\psi_t(p) \propto (E^2 - m^2)^{-\frac{1}{4}} e^{iEt}. \tag{3.103}$$

This result corresponds to the necessary form of a wave function at some given value of m and concentrated at some value of t, the analog of the Newton–Wigner wave function for a particle concentrated at a given point $\mathbf{x}$. A simple computation shows that

$$-i\left(\frac{\partial}{\partial E} - \frac{iE}{2p^2}\right)\psi_t(p) = t\psi_t(p). \tag{3.104}$$

Thus, the operator that appears in the expectation value in (3.97) at each value of m in the foliation induced by the change of variables (3.94) corresponds to the analog of the Newton–Wigner position operator for time, restricted to a given mass value.

Clearly, the Fourier transform of the function $\psi_{t_0}(p)$ of (3.103) (picking the localization point to be $t = t_0$) into the time domain by the kernel $\exp -iEt$ would not be localized in t, as for the Newton–Wigner problem in $\mathbf{x}$, and would therefore not form a viable quantum theory if, as we have assumed, the mass is concentrated at a fixed point.

We remark that, as for $\mathbf{x}_{NW}$, the Landau–Peierls operator t_{LP} is a constant of the free motion (as can be easily verified by computing their commutator with the free Hamiltonian). The (mean) intercepts of the virtual motions contained in the wave function, respectively to $t = 0$ and to $x = 0$ do not change under the free motion.

Chapter 4

Spin in Relativistic Quantum Theory

4.1 Definition of Spin

We start this chapter by discussing the notion of the *spin* of a particle in the framework of non-relativistic quantum theory, intrinsically associated with a manifestation of angular momentum. Classically, the angular momentum of a particle about some chosen origin is defined as

$$\mathbf{L} = \mathbf{r} \times m\mathbf{v} \equiv \mathbf{r} \times \mathbf{p}, \qquad (4.1)$$

or, in terms of components of the three-vectors,

$$L_i = x_j p_k - x_k p_j. \qquad (4.2)$$

With the operator relations (3.50), one finds the well-known commutation relations (summed on k)

$$[L_i, L_j] = i\hbar \epsilon_{ijk} L_k \qquad (4.3)$$

where ϵ_{ijk} is the totally antisymmetric tensor in three dimensions with values 0 or ± 1. The quantites (4.2) are the *generators* of the rotation group in three dimensions; the operator

$$U(\omega) = e^{i\hbar \omega \cdot \mathbf{L}} \qquad (4.4)$$

rotates the vector $\mathbf{r}$ (as well as $\mathbf{p}$) by an angle ω around the direction of the vector ω. If the Hamiltonian of the system H is invariant under such rotations, then $\mathbf{L}$ is conserved, *i.e.* constant in time.

We can see this by computing the expectation value of an observable $\mathcal{O}$ in a time dependent state $\psi_t = e^{iHt/\hbar}\psi$:

$$(\psi_t, \mathcal{O}\psi_t) = (\psi, e^{-iHt\hbar}\mathcal{O}e^{iHt/\hbar}\psi). \qquad (4.5)$$

If $[H, \mathcal{O}] = 0$, this expectation value will remain constant. Since $\mathbf{L}$ commutes with H, it is conserved.

Operators with the relation (4.3) have remarkable properties [e.g. Biedenharn and Louck (1981), Wigner (1931), Merzbacher (1970) and others]. The Schrödinger equation for a particle with a spherially symmetric potential (such as a Coulomb potential), after separation of variables has solutions called *spherical harmonics* for the angular dependence which exhibit these properties. Here, we demonstrate these properties entirely algebraically directly from the relations (4.3).

Let us define the operators

$$L_\pm = L_x, \pm iL_y. \qquad (4.6)$$

It then follows from (4.3) that

$$[L_+, L_-] = 2\hbar L_z \qquad (4.7)$$

and, furthermore,

$$\begin{aligned} [L_z, L_+] &= \hbar L_+ \\ [L_z, L_-] &= -\hbar L_- \end{aligned} \qquad (4.8)$$

These relations will enable us to generate the eigenvectors of the angular momentum operators. It follows, with the help of these algebraic relations, that

$$L_\pm L_\mp = \mathbf{L}^2 - L_z{}^2 \pm \hbar L_z \qquad (4.9)$$

We may now study the eigenfunctions of these operators. Since L_z commutes with $\mathbf{L}^2$ (they form a complete set of commuting operators), we can find a common set of eigenfunctions $\{|\lambda, m\rangle\}$, for which $\mathbf{L}^2$ has the eigenvalue λ and L_z the value m. We shall find that our algebra admits *both integer and half integer* eigenvalues, a mathematical result for which the half integral values actually occur in nature.

We now show that necessarily, $\lambda \geq m^2$. This follows from the fact that [Merzbacher (1970)]

$$\mathbf{L}^2 - L_z^{\,2} = L_x^{\,2} + L_y^{\,2} = \frac{1}{2}(L_+ L_+^\dagger + L_+^\dagger L_+) \geq 0, \qquad (4.10)$$

so that

$$\langle \lambda, m | \mathbf{L}^2 - L_z^{\,2} | \lambda, m \rangle \geq 0, \qquad (4.11)$$

i.e.,

$$\lambda \geq m^2. \qquad (4.12)$$

To calculate the spectrum, we now show that $L_\pm$ are *stepping operators*, in the sense that

$$L_z L_+ |\lambda, m\rangle = L_+ L_z |\lambda, m\rangle + \hbar L_+ |\lambda, m\rangle$$
$$= (m + 1)\hbar L_+ |\lambda, m\rangle, \qquad (4.13)$$

and

$$L_z L_- |\lambda, m\rangle = L_- L_z |\lambda, m\rangle - \hbar L_- |\lambda, m\rangle$$
$$= (m - 1)\hbar L_- |\lambda, m\rangle. \qquad (4.14)$$

Since $\mathbf{L}^2$ commutes with $L_\pm$, we may write

$$\mathbf{L}^2 L_\pm |\lambda, m\rangle = \lambda L_\pm |\lambda, m\rangle. \qquad (4.15)$$

With (4.12) and (4.13), we can write then that

$$L_\pm |\lambda, m\rangle = C_\pm(\lambda.m)\hbar |\lambda, m \pm 1\rangle. \qquad (4.16)$$

Stepping m to it greatest value, which we shall call ℓ, we must have

$$L_+ |\lambda, \ell\rangle = 0. \qquad (4.17)$$

We can now find λ in terms of ℓ by multiplying by L_-:

$$L_- L_+ |\lambda \ell\rangle = (\mathbf{L}^2 - L_z^{\,2} - \hbar L_z)|\lambda \ell\rangle$$
$$= (\lambda - \ell^2 - \ell)|\lambda \ell\rangle = 0, \qquad (4.18)$$

so that

$$\lambda = \ell(\ell + 1). \qquad (4.19)$$

Now, suppose the lowest value of m is ℓ'. Then,

$$L_-|\lambda\ell'\rangle = 0. \tag{4.20}$$

Multiplying by L_+, we find in the same way,

$$\lambda = \ell'(\ell' - 1). \tag{4.21}$$

For the same λ, we can have $\ell' = -\ell$ or $\ell' = \ell + 1$. The second choice is beyond our bound, so we must have

$$\ell' = -\ell. \tag{4.22}$$

In steps from ℓ to $-\ell$,

$$\ell - \ell' = 2\ell \tag{4.23}$$

must be a non-negative integer. Therefore,

$$\ell = 0, \frac{1}{2}, 1, \frac{3}{2}, \dots. \tag{4.24}$$

In the stepping operation, we can write the transition upwards and downwards in terms of a sequence of states,

$$L_\pm|\lambda, m\rangle = C_\pm(\lambda, m)\hbar|\lambda, m \pm 1\rangle, \tag{4.25}$$

so that, assuming the state functions normalized to unity,

$$\langle\lambda, m|L_-L_+|\lambda m\rangle = |C_+|^2\hbar^2 \tag{4.26}$$

and

$$\langle\lambda, m|L_+L_-|\lambda m\rangle = |C_-|^2\hbar^2. \tag{4.27}$$

These results are a direct inference from the commutation relations of the components of $\mathbf{L}$. The properties of this operator were modelled by the commutation relations of the components of $\mathbf{r} \times \mathbf{p}$, an *orbital angular momentum*. This operator occurs in a natural way in the solution of the spherically symmetric central force problem, such as the hydrogen atom. It is an identity that

$$\mathbf{L}^2 = (\mathbf{r} \times \mathbf{p}) \cdot (\mathbf{r} \times \mathbf{p}) = r^2\mathbf{p}^2 - \mathbf{r}(\mathbf{r} \cdot \mathbf{p}) \cdot \mathbf{p} + 2i\hbar(\mathbf{r} \cdot \mathbf{p}), \tag{4.28}$$

so that, after a little manipulation, we see that

$$\mathbf{L}^2 = r^2\mathbf{p}^2 + \hbar^2\frac{\partial}{\partial r}\left(r^2\frac{\partial}{\partial r}\right). \tag{4.29}$$

This relation gives us a simple connection between the Laplacian (proportional to $\mathbf{p}^2$) that occurs in the Schrödinger equation and the operator $\mathbf{L}^2$. Separating variables in spherical coordinates, the solutions for the angular part of the wave function are spherical harmonics indexed by ℓ, m, resulting in normalizable wave functions *only for ℓ integer*. Experimentally, a magnetic field $\mathbf{B}$ would interact with the atom in a form proportional to $\mathbf{L} \cdot \mathbf{B}$. For $\mathbf{B}$ defining the z direction, the induced energy shift would be proportional to m, so that there would be a splitting to $2\ell + 1$ levels. For ℓ integer, this would an *odd* number. It is observed, however, that there are cases for which there is a splitting to an *even* number of levels, implying the existence in nature of the *half integer* values of ℓ.

As a consequence of this observation, it was suggested by H. Kramers [Gottfried and Weisskopf (1986)] to W. Pauli that the electron has an additional internal degree of freedom, interacting with a magnetic field through an associated magnetic moment, which was called *spin*. Pauli initially rejected the idea, and Einstein [Gottfried and Weisskopf (1986)] showed that a rotating charged object could only have integer values of angular momentum. Uhlenbeck and Goudschmidt independently suggested, in Holland, the idea of spin to Lorentz, who reportedly (1926) said that they were young enough to write it in a paper [Gottfried and Weisskopf (1986)]. However, the analysis of L.H. Thomas of spin orbit coupling (1927) finally convinced Einstein that these representations exist in nature.[1]

The commutation relations that we wrote above constitute the infinitesimal algebra of the *rotation group* in three dimensions, as pointed out above (Eqs. (4.3), (4.4)). The half integer structure is the lowest non-trivial reprentation. In fact, it is the simplest non-trivial quantum system. As Schwinger (1951) has pointed out, one may

---

[1]It is not necessarily true that the lowest representation of a symmetry group found in nature would be manifested in the laboratory. The elements of the lowest representation of the classification group $SU(3)$ for elementary particles have been assumed to have an objective existence, called "quarks", with fruitful consequences (1961); however, they are not directly observed, and it is a consequence of quantum field theory that they cannot be seen directly, but nevertheless may have physical reality.

think of a quantum system with just two states, up represented by

$$\begin{pmatrix} 1 \\ 0 \end{pmatrix} \tag{4.30}$$

and down

$$\begin{pmatrix} 0 \\ 1 \end{pmatrix}. \tag{4.31}$$

The matrix operator

$$\begin{pmatrix} 1 & 0 \\ 0 & 0 \end{pmatrix} \tag{4.32}$$

$$\begin{pmatrix} 1 & 0 \\ 0 & -1 \end{pmatrix} \tag{4.33}$$

has eigenvalues ± 1 (up, down) on these two states.

The transition from down to up is achieved by multiplying by the matrix

$$\begin{pmatrix} 0 & 1 \\ 0 & 0 \end{pmatrix} \tag{4.34}$$

and from up to down by

$$\begin{pmatrix} 0 & 0 \\ 1 & 0 \end{pmatrix}. \tag{4.35}$$

These matrices are not Hermitian, However, the linear combinations

$$\sigma_1 = \begin{pmatrix} 0 & 1 \\ 1 & 0 \end{pmatrix} \tag{4.36}$$

and

$$\sigma_2 = \begin{pmatrix} 0 & -i \\ i & 0 \end{pmatrix} \tag{4.37}$$

are Hermitian and have the commutation relations

$$\left[\frac{\sigma_1}{2}, \frac{\sigma_2}{2} \right] = i \frac{\sigma_3}{2}, \tag{4.38}$$

and are cyclic, just the commutation relations (4.3) of angular momentum. This two state system constitutes the lowest non-trivial representation of the rotation group. It is precisely this operator, along with other half integer angular momenta, which can account for the even number of levels that atomic lines split.

We finally remark that, since each of the $\sigma_i{}^2 = 1$, a rotation of 2π results (due to the factor $\frac{1}{2}$ in (4.4)) in multiplication by -1 of the wave function. A rotation of 4π results in a factor of unity. Thus, we have a group that is called the *double covering* of the rotation group $O(3)$, de noted by $SU(2)$ (unitary with determinant one). [Jones (1998)].

The analysis we have so far presented has been completly non-relativistic, and does not take into account the effect of Lorentz transformations.

We shall discuss in the remainder of this chapter, the basic idea of a relativistic particle with spin, is based on Wigner's seminal work [Wigner (1939)].[2] As we shall show, Wigner's original construction in not consistent with a full relativistic quantum theory. His procedure is adapted here to construct a form which is applicable to such a relativistic quantum theory; in this form, Wigner's theory, together with the requirements imposed by the observed correlation between spin and statistics in nature for identical particle systems, makes it possible to define the total spin of a state of a relativistic many body system.

4.2 Relativistic Spin and the Dirac Representation

The spin of a particle in a nonrelativistic framework, as we have pointed out, corresponds to the lowest dimensional nontrivial representation of the rotation group; the generators are the Pauli matrices σ_i divided by two, the generators of the fundamental representation of the double covering of $SO(3)$. The self-adjoint operators that are the generators of this group measure angular momentum and are associated with magnetic moments.

---

[2]Dirac (1947) formulated a relativistic wave equation, to be discussed later, from which the spin emerges, in interaction with electromagnetism, in the nonrelativistic limit.

In the nonrelativistic quantum theory, the spin states of a two or more particle system is defined by combining the spins of these particles at equal time using appropriate Clebsch–Gordan coefficients [Clebsch (1872)] at each value of the time. The restriction to equal time follows from the tensor product form of the representation of the quantum states for a many body problem [Baym (1969), Fetter(1971)]. For two spin 1/2 (Fermi-Dirac) particles, an antisymmetric space distribution would correspond to a symmetric combination of the spin factors, *i.e.* a spin one state, and a symmetric space distribution would correspond to an antisymmetric spin combination, a spin zero state. This correlation is the source of the famous Einstein–Podolsky–Rosen discussion [Einstein (1935)] and provides an important model for quantum information transfer. The experiment proposed by Palacios *et al.* [Palacios (2009)]suggests that spin entanglement can occur for two particles at non-equal times; the spin carried by wave fnctions of SHP type would naturally carry such correlations over the width in t of the wave packets, and therefore the formulation we shall present here would be appropriate for application to relativistic quantum information transfer (*e.g.*,[Aharonov (1982), Hu (2012), Lin (2009), Lizier (2013)]]).

Wigner (1939) worked out a method for defining spin for relativistic particles. This formulation is not appropriate for application to quantum theory, since it does not preserve, as we shall explain below, the covariance of the expectation value of coordinate operators. Before constructing a generalization of Wigner's method which is useful in relativistic quantum theory we first review Wigner's method in its original form, and show how the difficulties arise.

The method developed by Wigner provided the foundation for what is now known as the theory of induced representations [Mackey (1968)], with very wide applications, including a very powerful approach to finding the representations of noncompact groups.

We shall show here how Wigner's approach can be used to describe the spin of a particle in the framework of the manifestly covariant consistent quantum theory of Stueckelberg, Horwitz and Piron [SHP; Stueckelberg (1941), Horwitz (1973)], and how this method can be extended to describe the combined spin states of a many body system.

To establish some notation and the basic method, we start with the basic principle of relativistic covariance for a scalar quantum wave function $\psi(p)$. In a new Lorentz frame described by the parameters Λ of the Lorentz group, for which $p'^\mu = \Lambda^\mu_\nu p^\nu$ (we work in momentum space here for convenience), the same physical point in momentum space, described in a different frame, must have the same probability density, *i.e.*,

$$\psi'(p') = \psi(p) \tag{4.39}$$

up to a phase, which we take to be unity. It then follows that as a function of p,

$$\psi'(p) = \psi(\Lambda^{-1}p). \tag{4.40}$$

Since, in Dirac's notation (1947),

$$\psi'(p) \equiv \langle p|\psi'\rangle, \tag{4.41}$$

Eq. (4.40) follows equivalently by writing

$$|\psi'\rangle = U(\Lambda)|\psi\rangle \tag{4.42}$$

so that

$$\begin{aligned}
\langle p|\psi'\rangle &= \langle p|U(\Lambda)|\psi\rangle \\
&= \langle \Lambda^{-1}p|\psi\rangle \\
&= \psi(\Lambda^{-1}p),
\end{aligned} \tag{4.43}$$

where we have used

$$U(\Lambda)^\dagger|p\rangle = U(\Lambda^{-1})|p\rangle = |\Lambda^{-1}p\rangle.$$

To discuss the transformation properties of the representation of a relativistic particle with spin, Wigner (1937) proposed that we consider a special frame in which $p_0^\mu = (m, 0, 0, 0)$; the subgroup of the Lorentz group that leaves this vector invariant is clearly $O(3)$, the rotations in the three space in which $\mathbf{p} = 0$, or its covering $SU(2)$. Under a Lorentz boost, transforming the system to its representation in a moving inertial frame, the rest momentum appears as $p_0^\mu \to p^\mu$, but under this unitary transformation, the subgroup that leaves p_0^μ invariant is carried to a form which leaves p^μ invariant, and the group remains $SU(2)$. The 2×2 matrices representing this group are altered by the Lorentz transformation, and are functions of the momentum p^μ. The resulting state then transforms by a further change in p^μ and an $SU(2)$ transformation compensating for this change. This additional transformation is called the "little group" of Wigner. The family of values of p^μ generated by Lorentz transformations on p_0^μ is called the "orbit" of the induced representation. This $SU(2)$, in its lowest dimensional representation, parametrized by p^μ and the additonal Lorentz transformation Λ, corresponds to Wigner's covariant relativistic definition of the spin of a relativistic particle [Wigner (1937)].

We now apply this method to review Wigner's construction based on a representation induced on the momentum p^μ. Let us *define* the momentum-spin ket

$$|p, \sigma\rangle \equiv U(L(p))|p_0, \sigma\rangle, \qquad (4.44)$$

where $U(L(p))$ is the unitary operator inducing a Lorentz transformation of the timelike $p_0 = (m, 0, 0, 0)$ (rest frame momentum) to the general timelike vector p^μ. The effect of a further Lorentz transformation parameterized by Λ, induced by $U(\Lambda^{-1})$, can be written as

$$U(\Lambda^{-1})|p, \sigma\rangle = U(L(\Lambda^{-1}p))U^{-1}(L(\Lambda^{-1}p))U(\Lambda^{-1})U(L(p))|p_0, \sigma\rangle$$
$$(4.45)$$

The product of the last three unitary factors

$$U^{-1}(L(\Lambda^{-1}p))U(\Lambda^{-1})U(L(p)) \qquad (4.46)$$

has the property that under this combined unitary transformation, the ket is transformed so that $p_0 \to p_0$, and thus corresponds to just a rotation (called the Wigner rotation), the stability subgroup of the vector p_0. This rotation can be represented by a 2×2 matrix acting on the index σ, *i.e.*, so that

$$U(\Lambda^{-1})|p, \sigma\rangle = U(L(\Lambda^{-1}p))|p_0, \sigma'\rangle D_{\sigma, \sigma'}(\Lambda, p)$$
$$= |\Lambda^{-1}p, \sigma'\rangle D_{\sigma, \sigma'}(\Lambda, p). \qquad (4.47)$$

where, as a representation of rotations, D is unitary. Therefore, taking the complex conjugate of

$$\langle \psi | U(\Lambda^{-1}) | p, \sigma \rangle = \langle \psi | \Lambda^{-1} p, \sigma' \rangle D_{\sigma, \sigma'}(\Lambda, p),$$

one obtains $(A^{\dagger *} = A^T)$

$$\langle p, \sigma | U(\Lambda) \psi \rangle = \langle \Lambda^{-1} p, \sigma' | \psi \rangle D_{\sigma', \sigma}(\Lambda p), \tag{4.48}$$

where, in this construction,

$$D_{\sigma', \sigma}(\Lambda, p) = \left((L(p)^{-1} \Lambda L(\Lambda^{-1} p)) \right)_{\sigma', \sigma}, \tag{4.49}$$

expressed in terms of the $SL(2, C)$ matrices corresponding to the unitary transformation with determinant unity (4.46).

The result (4.48) can be written as

$$\psi'(p, \sigma) = \psi(\Lambda^{-1} p, \sigma') D_{\sigma', \sigma}(\Lambda, p), \tag{4.50}$$

in accordance with (2.68), generalized to take into account the spin degrees of freedom of the wavefunction. The algebra of the 2×2 matrices of the fundamental representation of the group $SL(2, C)$ are isomorphic to that of the Lorentz group, and the product of the corresponding matrices provide the 2×2 matrix representation of $D_{\sigma', \sigma}(\Lambda, p)$; we may therefore write (4.49) as

$$D_{\sigma', \sigma}(\Lambda, p) = \left(L^{-1}(p) \Lambda L(\Lambda^{-1} p) \right)_{\sigma', \sigma}, \tag{4.51}$$

where L and Λ are the 2×2 matrices of $SL(2, C)$.

As we have mentioned above, the presence of the p-dependent matrices representating the spin of a relativistic particle in the transformation law of the wave function destroys the covariance, in a relativistic quantum theory, of the expectation value of the coordinate operators. To see this, consider the expectation value of the dynamical variable x^μ, *i.e.*

$$\langle x^\mu \rangle = \int d^4 p \, \psi(p)^\dagger i \frac{\partial}{\partial p_\mu} \psi(p). \tag{4.52}$$

A Lorentz transformation would introduce the p-dependent 2×2 unitary transformation on the function $\psi(p)$, and the derivative with respect to momentum would destroy the covariance property that we would wish to see of the expectation value $\langle x^\mu \rangle$.

It is also not possible, in this framework, to form wave packets of definite spin by integrating over the momentum variable, since this would add functions over different parts of the orbit, with a different $SU(2)$ at each point.

As will be described in the following, these problems were solved by inducing a representation of the spin on a timelike unit vector n^μ in place of the four-momentum, using a representation induced on a timelike vector, say, n^μ, which is independent of x^μ or p^μ [Horwitz (1975), Arshansky (1982)]. This solution also permits the linear superposition of momentum states to form wave packets of definite spin, and admits the construction of definite spin states for many body relativistic systems and its consequences for entanglement. In the following, we show how such a representation can be constructed, and discuss some of its dynamical implications.

To carry out this program, let us define, as in (4.44),

$$|n, \sigma, x\rangle \equiv U(L(n))|n_0, \sigma, x\rangle, \tag{4.53}$$

where we may admit a dependence on x (or, through Fourier transform, on p). Here, we distinguish the action of $U(L(n))$ from the general Lorentz transformation $U(\Lambda)$; $U(L(n))$ acts only on the vector space of the n^μ. Its infinitesimal generators are given by

$$M_n^{\mu\nu} = -i\left(n^\mu \frac{\partial}{\partial n_\nu} - n^\nu \frac{\partial}{\partial n_\mu}\right), \tag{4.54}$$

while the generators of the transformations $U(\Lambda)$ act on the full vector space of both the n^μ and the x^μ (as well as p^μ). In terms of the canonical variables,

$$M^{\mu\nu} = M_n^{\mu\nu} + (x^\mu p^\nu - x^\nu p^\mu). \tag{4.55}$$

The two terms of the full generator commute. Following the method outlined above, we now investigate the properties of a total Lorentz transformation, *i.e.*,

$$U(\Lambda^{-1})|n, \sigma, x\rangle = U(L(\Lambda^{-1}n)\big(U^{-1}(L(\Lambda^{-1}n))U(\Lambda^{-1})$$
$$\times U(L(n))\big)\big)n_0, \sigma, x\rangle, \tag{4.56}$$

Now, consider the conjugate of (4.56),

$$\langle n, \sigma, x | U(\Lambda) = \langle n_0, \sigma, x | \left(U(L^{-1}(n)) U(\Lambda) U(L(\Lambda^{-1}n)) \right)$$
$$\times U^{-1}(L(\Lambda^{-1}n)). \tag{4.57}$$

The operator in the first factor (in parentheses) preserves n_0, and therefore, as before, contains an element of the little group associated with n^μ which may be represented by the matrices of $SL(2, C)$. It also acts, due to the factor $U(\Lambda)$ (for which the generators are those of the Lorentz group acting both on n and x (or p), as in (4.55)), taking $x \to \Lambda^{-1}x$ in the conjugate ket on the left. Taking the product on both sides with $|\psi\rangle$, we obtain

$$\langle n, \sigma, x | \psi \rangle' = \langle \Lambda^{-1}n, \sigma', \Lambda^{-1}x | \psi \rangle D_{\sigma',\sigma}(\Lambda, n), \tag{4.58}$$

or

$$\psi'_{n,\sigma}(x) = \psi_{\Lambda^{-1}n,\sigma'}(\Lambda^{-1}x) D_{\sigma',\sigma}(\Lambda, n), \tag{4.59}$$

where

$$D(\Lambda, n) = L^{-1}(n)\Lambda L(\Lambda^{-1}n), \tag{4.60}$$

with Λ and $L(n)$ the corresponding 2×2 matrices of $SL(2, C)$. Λ and $L(n)$ to be the corresponding 2×2 matrices of $SL(2, C)$.

With this transformation law, one may take the Fourier transform to obtain the wave function in momentum space, and conversely. The matrix D is an element of $SU(2)$, and therefore linear superpositions over momenta or coordinates maintain the definition of the particle spin, and interference phenomena for relativistic particles with spin may be studied consistently. Furthermore, if two or more particles with spin are represented in representations induced on n^μ, at a given value of n^μ on their respective orbits, their spins can be added by the standard methods with the use of Clebsch–Gordan coefficients [Clebsch (1872)]. This method therefore admits the treatment of a many body relativistic system with spin.

Our assertion of the unitarity of the n-dependent part of the transformation has assumed that the integral measure on the Hilbert space, to admit integration by parts, is of the form $d^4n\,d^4x\,\delta(n^\mu n_\mu + 1)$, *i.e.*, although the timelike vector n^μ, in many applications, is degenerate, it carries a probability interpretation under the norm, and may play a dynamical role.

There are two fundamental representations of $SL(2, C)$ which are inequivalent [Boerner (1963)]. Multiplication by the operator $\sigma \cdot p$ of a two dimensional spinor representating one of these results in an object transforming like the second representation. Such an operator could be expected to occur in a dynamical theory, and therefore the state of lowest dimension in spinor indices of a physical system should contain both representations.

We now discuss the construction of Dirac spinors [Dirac (1947)]. An approximate treatment of the Dirac equation in interaction with electromagnetism, yields a connection with spin, identified through its interaction with the magnetic field [Bjorken (1964)]. As we shall see, however, the particle spin is already contained in the construction of the Dirac function through the fundamental construction of Wigner, combining the two fundamental representations of $SL(2, C)$ [Arshansky (1982), Weinberg (1995)].

We first remark that the defining relation for the fundamental $SL(2, C)$ matrices is

$$\Lambda^\dagger \sigma^\mu n_\mu \Lambda = \sigma^\mu (\Lambda^{-1} n)_\mu, \tag{4.61}$$

where $\sigma^\mu = (\sigma^0, \sigma)$; σ^0 is the unit 2×2 matrix, and σ are the Pauli matrices. Since the determinant of $\sigma^\mu n_\mu$ is the Lorentz invariant $n^{0^2} - \mathbf{n}^2$, and the determinant of Λ is unity in $SL(2, C)$, the transformation represented on the left hand side of (4.61) must induce a Lorentz transformation on n^μ. An inequivalent second fundamental representation may be constructed by using this defining relation with σ^μ replaced by $\underline{\sigma}^\mu \equiv (\sigma^0, -\sigma)$. For every Lorentz transformation Λ acting on n^μ, this defines an $SL(2, C)$ matrix $\underline{\Lambda}$ (we use the same symbol for the Lorentz transformation on a four-vector as for the corresponding $SL(2, C)$ matrix acting on the 2-spinors).

Since both fundamental representations of $SL(2, C)$ should occur in the general quantum wave function representing the state of the system, the norm for each n-sector of the Hilbert space must be defined as

$$N = \int d^4 x (|\hat{\psi}_n(x)|^2 + |\hat{\phi}_n(x)|^2), \tag{4.62}$$

where $\hat{\psi}_n$ transforms with the first $SL(2,C)$ and $\hat{\phi}_n$ with the second. From the construction of the little group (4.60), it follows that $L(n)\psi_n$ transforms with Λ, and $\underline{L}(n)\phi_n$ transforms with $\underline{\Lambda}$; making this replacement in (4.62), and using the fact, obtained from the defining relation (4.61), that $L(n)^{\dagger-1}L(n)^{-1} = \mp\sigma^\mu n_\mu$ and $\underline{L}(n)^{\dagger-1}\underline{L}(n)^{-1} = \mp\underline{\sigma}^\mu n_\mu$, one finds that

$$N = \mp \int d^4x \bar{\psi}_n(x)\gamma \cdot n\psi_n(x), \qquad (4.63)$$

where $\gamma \cdot n \equiv \gamma^\mu n_\mu$ (for which $(\gamma \cdot n)^2 = -1$), and the matrices γ^μ are the Dirac matrices as defined in the books of Bjorken and Drell [Bjorken (1964)]. Here, the four-spinor $\psi_n(x)$ is defined by [Weinberg (1995)]

$$\psi_n(x) = \frac{1}{\sqrt{2}}\begin{pmatrix} 1 & 1 \\ -1 & 1 \end{pmatrix}\begin{pmatrix} L(n)\hat{\psi}_n(x) \\ \underline{L}(n)\hat{\phi}_n(x) \end{pmatrix}, \qquad (4.64)$$

and the sign $\mp$ corresponds to n^μ in the positive or negative light cone. The wave function transforms as

$$\psi'_n(x) = S(\Lambda)\psi_{\Lambda^{-1}n}(\Lambda^{-1}x) \qquad (4.65)$$

and $S(\Lambda)$ is a (nonunitary) transformation generated infinitesimally, as in the standard Dirac theory (see, for example, [Bjorken (1964), Weinberg (1995)]), by $\Sigma^{\mu\nu} \equiv \frac{i}{4}[\gamma^\mu, \gamma^\nu]$.

The Dirac operator $\gamma \cdot p$ is not Hermitian in the (invariant) scalar product associated with the norm (4.63). It is of interest to consider the Hermitian and anti-Hermitian parts

$$K_L = \frac{1}{2}(\gamma \cdot p + \gamma \cdot n\gamma \cdot p\gamma \cdot n) = -(p \cdot n)(\gamma \cdot n)$$

$$K_T = \frac{1}{2}\gamma^5(\gamma \cdot p - \gamma \cdot n\gamma \cdot p\gamma \cdot n) = -2i\gamma^5(p \cdot K)(\gamma \cdot n), \qquad (4.66)$$

where $K^\mu = \Sigma^{\mu\nu}n_\nu$, and we have introduced the factor $\gamma^5 = i\gamma^0\gamma^1\gamma^2\gamma^3$, which anticommutes with each γ^μ and has square -1 so that K_T is Hermitian and commutes with the Hermitian K_L. Since

$$K_L^2 = (p \cdot n)^2 \qquad (4.67)$$

and

$$K_T^2 = p^2 + (p \cdot n)^2, \tag{4.68}$$

we may consider

$$K_T^2 - K_L^2 = p^2 \tag{4.69}$$

to pose an eigenvalue problem analogous to the second order mass eigenvalue condition for the free Dirac equation (the Klein Gordon condition). For the Stueckelberg equation of evolution corresponding to the free particle, we may therefore take

$$K_0 = \frac{1}{2M}(K_T^2 - K_L^2) = \frac{1}{2M}p^2. \tag{4.70}$$

In the presence of electromagnetic interaction, gauge invariance under a spacetime dependent gauge transformation requires that the expressions for K_T and K_L given in (4.66) should be, in place of (4.70), the gauge invarint form

$$K = \frac{1}{2M}(p - eA)^2 + \frac{e}{2M}\Sigma_n^{\mu\nu} F_{\mu\nu}(x), \tag{4.71}$$

where

$$\Sigma_n^{\mu\nu} = \Sigma^{\mu\nu} + K^\mu n^\nu - K^\nu n^\mu \equiv \frac{i}{4}[\gamma_n^\mu, \gamma_n^\nu], \tag{4.72}$$

and the γ_n^μ are defined below in (4.76). The expression (4.71) is quite similar to that of the second order Dirac operator; it is, however, Hermitian and has no direct electric coupling to the electromagnetic field in the special frame for which $n^\mu = (1,0,0,0)$ in the minimal coupling model we have given here (note that in his calculation of the anomalous magnetic moment [Schwinger (1951)], Schwinger puts the electric field to zero; a non-zero electric field would lead to a non-Hermitian term in the standard Dirac propagator, the inverse of the Klein-Gordon square of the interacting Dirac equation). The matrices $\Sigma_n^{\mu\nu}$ are, in fact, a relativistically covariant form of the Pauli matrices.

To see this, we note that the quantities K^μ and $\Sigma_n^{\mu\nu}$ satisfy the commutation relations

$$[K^\mu, K^\nu] = -i\Sigma_n^{\mu\nu}$$

$$[\Sigma_n^{\mu\nu}, K^\lambda] = -i[(g^{\mu\lambda} + n^\nu n^\lambda)K^\mu - (g^{\mu\lambda} + n^\mu n^\lambda)K^\nu,$$

$$[\Sigma_n^{\mu\nu}, \Sigma_n^{\lambda\sigma}] = -i[(g^{\nu\lambda} + n^\nu n^\lambda)\Sigma_n^{\mu\sigma} + (g^{\sigma\mu} + n^\sigma n^\mu)\Sigma_n^{\lambda\nu}$$

$$- (g^{\mu\lambda} + n^\mu n^\lambda)\Sigma_n^{\nu\sigma} + (g^{\sigma\nu} + n^\sigma n^\nu)\Sigma_n^{\lambda\mu}]. \qquad (4.73)$$

Since $K^\mu n_\mu = n_\mu \Sigma_n^{\mu\nu} = 0$, there are only three independent K^μ and three $\Sigma_n^{\mu\nu}$. The matrices $\Sigma_n^{\mu\nu}$ are a covariant form of the Pauli matrices, and the last of (4.73) is the Lie algebra of $SU(2)$ in the spacelike surface orthogonal to n^μ. The three independent K^μ correspond to the non-compact part of the algebra which, along with the $\Sigma_n^{\mu\nu}$ provide a representation of the Lie algebra of the full Lorentz group. The covariance of this representation follows from

$$S^{-1}(\Lambda)\Sigma_{\Lambda n}^{\mu\nu} S(\Lambda)\Lambda_\mu^\lambda \Lambda_\nu^\sigma = \Sigma_n^{\lambda\sigma}. \qquad (4.74)$$

In the special frame for which $n^\mu = (1, 0, 0, 0)$, $\Sigma_n^{i,j}$ become the Pauli matrices $\frac{1}{2}\sigma^k$ with (i, j, k) cyclic, and $\Sigma_n^{0j} = 0$. In this frame there is no direct electric interaction with the spin in the minimal coupling model (4.72). We remark that there is, however, a natural spin coupling which becomes pure electric in the special frame, given by

$$i[K_T, K_L] = -ie\gamma^5(K^\mu n^\nu - K^\nu n^\mu)F_{\mu\nu}. \qquad (4.75)$$

It is a simple exercise to show that the value of this commutator reduces to $\mp e\sigma \cdot \mathbf{E}$ in the special frame for which $n^0 = -1$; this operator is Hermitian and would correspond to an electric dipole interaction with the spin.

Note that the matrices

$$\gamma_n^\mu = \gamma_\lambda \pi^{\lambda\mu}, \qquad (4.76)$$

where the projection

$$\pi^{\lambda\mu} = g^{\lambda\mu} + n^\lambda n^\mu, \qquad (4.77)$$

appearing in (4.73), plays an important role in the description of the dynamics in the induced representation. In (4.71), the existence of projections on each index in the spin coupling term implies that $F^{\mu\nu}$ can be replaced by $F_n{}^{\mu\nu}$ in this term, a tensor projected into the foliation subspace.

We further remark that in relativistic scattering theory, the S-matrix is Lorentz invariant [Bjorken (1964)]. The asymptotic states can be decomposed according to the conserved projection operators

$$P_\pm = \frac{1}{2}(1 \mp \gamma \cdot n)$$

$$P_{E\pm} = \frac{1}{2}\left(1 \mp \frac{p \cdot n}{|p \cdot n|}\right)$$

and

$$P_{n\pm} = \frac{1}{2}\left(1 \pm \frac{2i\gamma^5 K \cdot p}{[p^2 + (p \cdot n)^2]^{1/2}}\right). \tag{4.78}$$

The operator

$$\frac{2i\gamma^5 K \cdot p}{[p^2 + (p \cdot n)^2]^{1/2}} \to \sigma \cdot \mathbf{p}/|\mathbf{p}|; \tag{4.79}$$

when $n^\mu \to (1,0,0,0)$, $P_{n\pm}$ corresponds to a helicity projection.Therefore the matrix elements of the S-matrix at any point on the orbit of the induced representation is equivalent (by replacing S by $U(L(n))SU^{-1}(L(n))$) to the corresponging helicity representation associated with the frame in which n is n_0.[3]

Note that the discrete symmetries act on the wave functions as

$$\psi_{\tau n}^C = C\gamma^0\psi_{-\tau n}^*(x)$$

$$\psi_{\tau n}^P(x) = \gamma^0\psi_{\tau, -\mathbf{n}, n^0}(-\mathbf{x}, t),$$

$$\psi_{\tau n}^T = i\gamma^1\gamma^3\psi_{-\tau, \mathbf{n}, -n^0}^*(\mathbf{x}, -t),$$

$$\psi_{\tau n}^{CPT}(x) = i\gamma^5\psi_{\tau, -n}(-\mathbf{x}, -t), \tag{4.80}$$

---

[3]This result is consistent with the suggestion of Y. Aharonov (1983) that n_0 may be interpreted as corresponding to the frame of the Stern-Gerlach apparatus in which the spin state is prepared.

where $C = i\gamma^2\gamma^0$. The CPT conjugate wavefunction, according to its evolution in τ, moves backwards in spacetime relative to the motion of $\psi_{\tau n}$. For a wave packet with $E < 0$ components, which moves backwards in t as τ goes forward, it is the CPT conjugate wave function which moves forward with charge $-e$, i.e., the observed antiparticle. No Dirac sea [Dirac (1932)] is required for the consistency of the theory, since unbounded transitions to $E < 0$ are prevented by conservation of K.

4.3 The Many Body Problem with Spin, and Spin-Statistics

As in the nonrelativistic quantum theory, one represents the state of an N-body system in terms of a basis given by the tensor product of N one-particle states, each an element of a one-particle Hilbert space. The general state of such an N-body system is given by a linear superposition over this basis [*e.g.* Fetter and Walecka (1971)]. Second quantization then corresponds to the construction of a Fock space, for which the set of all N body states, for all N are imbedded in a large Hilbert space, for which operators that change the number N are defined [Baym (1969)]. We shall discuss this structure in this section, and show, with our discussion of the relativistic spin given in the previous section, that the spin of a relativistic many-body system can be well-defined (see also, [Bennett (2015)]). In order to construct the tensor product space corresponding to the many-body system, we consider, as for the nonrelativistic theory, the product of wave functions which are elements of the same Hilbert space. In the nonrelativistic theory, this corresponds to functions at equal time; in the relativistic theory, the functions are taken to be at equal τ. Thus, in the relativistic theory, there are correlations at unequal t, within the support of the Stueckelberg wave functions. Moreover, for particles with spin we argue that in the induced representation, these functions must be taken at *identical values of n^μ*, i.e., taken at the same point on the orbits of the induced representation of each particle [Horwitz (2013)]:

Identical particles must be represented in tensor product states by wave functions at equal τ and equal n^μ.

The proof of this statement lies in the observation that the spin-statistics relation appears to be a universal fact of nature. The elementary proof of this statement, for example, for a system of two spin $1/2$ particles, is that a π rotation of the system introduces a phase factor of $e^{i\frac{\pi}{2}}$ for each particle, thus introducing a minus sign for the two body state. However, the π rotation is equivalent to an interchange of the two identical particles. This argument rests on the fact that each particle is in the same representation of $SU(2)$, which can only be achieved in the induced representation with the particles at the same point on their respective orbits. The same argument applies for bosons, which must be symmetric under interchange (in this case the phase of each factor in a pair is $e^{i\pi}$). We therefore see that identical particles must carry the same value of n^μ, and the construction of the N–body system must follow this rule. It therefore follows that the two body relativistic system can carry a spin computed by use of the usual Clebsch–Gordan coefficients, and entanglement would follow even at unequal time (within the support of the equal τ wave functions), as in the proposed experiment of Palacios *et al.* [Palacios (2009)]. This argument can be followed for arbitrary N, and therefore the Fock space of quantum field theory, as we show below, carries the properties usually associated with fermion (or boson) fields, with the entire Fock space foliated over the orbit of the inducing vector n^μ.

We remark that since the relativistic S-matrix is Lorentz invariant, the matrix elements of the S-matrix in states labelled by the asymptotic projections $P_{n\pm}$ (defined in (3.39)) can be replaced (by the substitution $U(L(n))SU^{-1}(L(n))$ for S) by helicities in the common frame in which $n^\mu \to (1,0,0,0)$. The Lorentz transformation that achieves this acts in the same way on all of the momenta of the asymptotic states and the resulting measured cross sections for this helicity representation then correspond to a choice of frame in which the common orbit is specified to be at the point $n^\mu = (1,0,0,0)$.

Although, due to the Newton–Wigner problem discussed above, the solutions of the Dirac equation are not suitable for the covariant local description of a quantum theory, the functions constructed in (4.64), under the norm (4.63), can form the basis of a consistent covariant quantum theory; they describe the (off-shell) states of a *local* quantum theory.

We then start by constructing a two body Hilbert space in the framework of the relativistic quantum theory. The states of this two body space are given by linear combinations over the product wave functions, where the wave functions for the spin $(1/2)$ case are given by the Dirac function of the type described in (4.64) (as well for integer spin functions) of the form

$$\psi_{ij}(x_1, x_2) = \psi_i(x_1) \times \psi_j(x_2), \qquad (4.81)$$

where $\psi_i(x_1)$ and $\psi_j(x_2)$ are elements of the one-particle Hilbert space $\mathcal{H}$. Let us introduce the notation, often used in differential geometry, that

$$\psi_{ij}(x_1, x_2) = \psi_i \otimes \psi_j(x_1, x_2), \qquad (4.82)$$

identifying the arguments according to a standard ordering. Then, without specifying the spacetime coordinates, we can write

$$\psi_{ij} = \psi_i \otimes \psi_j, \qquad (4.83)$$

formally, an element of the tensor product space $\mathcal{H}_1 \otimes \mathcal{H}_2$. The scalar product is carried out by pairing the elements in the two factors according to their order, since it corresponds to integrals over x_1, x_2, *i.e.*,

$$(\psi_{ij}, \psi_{k,\ell}) = (\psi_i, \psi_k)(\psi_j, \psi_\ell). \qquad (4.84)$$

For two identical particle states satisfying Bose–Einstein of Fermi–Dirac statistics, we must write, according to our argument given above,

$$\psi_{ijn} = \frac{1}{\sqrt{2}}[\psi_{in} \otimes \psi_{jn} \pm \psi_{jn} \otimes \psi_{in}], \qquad (4.85)$$

where $n \equiv n^u$ is the timelike four vector labelling the orbit of the induced representation. This expression has the required symmetry or antisymmetry only if both functions are on the same points of their respective orbits in the induced representation. Furthermore, they transform under the *same* $SU(2)$ representation of the rotation subgroup of the Lorentz group, and thus for spin $1/2$ particles, under a π spatial rotation (defined by the space orthogonal to the timelike vector n^μ) they both develop a phase factor $e^{i\frac{\pi}{2}}$. The product results in an over all negative sign. As in the usual quantum theory, this rotation corresponds to an interchange of the two particles, but here with respect to a "spatial" rotation around the vector n^μ. The spacetime coordinates in the functions are rotated in this (foliated) subspace of spacetime, and correspond to an actual exchange of the positions of the particles in space time, as in the formulation of the standard spin-statistics theorem. It therefore follows that the interchange of the particles occurs in the foliated space defined by n^μ, and, furthermore:

The antisymmetry of identical spin 1/2 (fermionic) particles remains at unequal times (within the support of the wave functions). This is true for the symmetry of identical spin zero (bosonic) particles as well.

The construction we have given enables us to define the spin of a many body system, even if the particles are relativistic and moving arbitrarily with respect to each other.

The spin of an N-body system is well-defined, independent of the state of motion of the particles of the system, by the usual laws of combining representations of $SU(2)$, i.e, with the usual Clebsch-Gordan coefficients, if the states of all the particles in the system are in induced representations at the same point of the orbit n^μ.

Thus, in the quark model for hadrons [Gell–Mann (1962), Ne'eman (1961)], the total spin of the hadron can be computed from the spins (and orbital angular momenta projected into the foliated space) of the individual quarks using the usual Clebsch–Gordan coefficients even if they are in significant relative motion, as part of the same $SU(2)$).

This result has important implications for the construction of the exchange interaction in many-body systems. Since there is no extra phase (corresponding to integer representations of the $SU(2)$) for the Bose–Einstein case, the boson symmetry can then be extended to a covariant symmetry with important implications for Bose–Einstein condensation.

4.4 Construction of the Fock Space and Quantum Field Theory

In the course of our construction, we have seen in detail that the foliation of the spacetime follows from the arguments based in the representations of a relativistic particle with half-integer spin. However, our considerations of the nature of identical particles, and their association with the spin statistics properties observed in nature, require that the foliation persists in the bosonic sector as well, where a definite phase $(\mp)$ under π rotations, exchanging two particles, must be in a definite representation of the rotation group specified by the foliation vector n^μ. We remark in this connection that the Cooper pairing [Cooper (1956)] of superconductivity must be between electrons on the same point of their induced represenation orbits, so that the superconducting state is defined on the corresponding foliation of spacetime as well. The resulting (quasi-) bosons have the identical particle properties inferred from our discussion of the boson sector.

The N body state of Fermi–Dirac particles can then be written as (the N body boson system should be treated separately since the normalization conditions are different, but we give the general result below)

$$\Psi_{nN} = \frac{1}{N!}\Sigma(-)^P P\psi_{nN} \otimes \psi_{nN-1} \otimes \cdots \psi_{n1}, \qquad (4.86)$$

where the permutations P are taken over all possibilities, and no two functions are equal. By the arguments given above, any pair of particle states in this set of particles have the Fermi–Dirac properties. We may now think of such a function as an element of a larger Hilbert space, called the *Fock space* which contains all values of the number N. On this space, one can define an operator that adds another particle (by multiplication), performs the necessary antisymmetrization, and changes the normalization appropriately. This operator is called a *creation operator*, which we shall denote by $a^\dagger(\psi_{nN+1})$ and has the property that

$$a^\dagger(\psi_{nN+1})\Psi_{nN} = \Psi_{nN+1}, \qquad (4.87)$$

now to be evaluated on the manifold $(x_{N+1}, x_N, x_{N-1} \ldots x_1)$. Taking the scalar product with some $N+1$ particle state Φ_{nN+1} in the Fock space, we see that

$$(\Phi_{nN+1}, a^\dagger(\psi_{nN+1})\Psi_{nN}) \equiv (a(\psi_{nN+1})\Phi_{nN+1}, \Psi_{nN}), \qquad (4.88)$$

thus defining the *annihilation* operator $a^\dagger(\psi_{nN+1})$.

The existence of such an annihilation operator, as in the usual construction of the Fock space, (*e.g.*, [Baym (1969)]) implies the existence of an additional element in the Fock space, the *vacuum*, or the state of no particles. The vacuum defined in this way lies in the foliation labelled by n^μ. The covariance of the construction, however, implies that, since all sectors labelled by n^μ are connected by the action of the Lorentz group, that this vacuum is an absolute vacuum for any n^μ, *i.e.*, the vacuum $\{\Psi_{n0}\}$ over all n^μ is Lorentz invariant.

The commutation relations of the annihilation-creation operators can be easily deduced from a low dimensional example, following the method used in the nonrelativistic quantum theory. Consider the two body state (4.83), and apply the creation operator $a^\dagger(\psi_{n3})$ to create the three body state

$$\Psi(\psi_{n3}, \psi_{n2}, \psi_{n1}) = \frac{1}{\sqrt{3!}} \{\psi_{n3} \otimes \psi_{n2} \otimes \psi_{n1} + \psi_{n1} \otimes \psi_{n3} \otimes \psi_{n2}$$

$$+ \psi_{n2} \otimes \psi_{n1} \otimes \psi_{n3} - \psi_{n2} \otimes \psi_{n3} \otimes \psi_{n1}$$

$$- \psi_{n1} \otimes \psi_{n2} \otimes \psi_{n3} - \psi_{n3} \otimes \psi_{n1} \otimes \psi_{n2}\} \quad (4.89)$$

One then takes the scalar product with the three body state

$$\Phi(\phi_{n3}, \phi_{n2}, \phi_{n1}) = \frac{1}{\sqrt{3!}} \{\phi_{n3} \otimes \phi_{n2} \otimes \phi_{n1} + \phi_{n1} \otimes \phi_{n3} \otimes \phi_{n2}$$

$$+ \phi_{n2} \otimes \phi_{n1} \otimes \phi_{n3} - \phi_{n2} \otimes \phi_{n3} \otimes \phi_{n1}$$

$$- \phi_{n1} \otimes \phi_{n2} \otimes \phi_{n3} - \phi_{n3} \otimes \phi_{n1} \otimes \phi_{n2}\} \quad (4.90)$$

Carrying out the scalar product term by term, and picking out the terms corresponding to scalar products of some functions with the two body state

$$\frac{1}{\sqrt{2}}\{\psi_{n2} \otimes \psi_{n1} - \psi_{n1} \otimes \psi_{n2}\} \tag{4.91}$$

one finds that the action of the adjoint operator $a(\psi_{n3})$ on the state $\Phi(\phi_{n3}, \phi_{n2}, \phi_{n1})$ is given by

$$a(\psi_{n3})\Phi(\phi_{n3}, \phi_{n2}, \phi_{n1}) = (\psi_{n3}, \phi_{n3})\phi_{n2} \otimes \phi_{n1} - (\psi_{n3}, \phi_{n2})\phi_{n3} \otimes \phi_{n1}$$
$$+ (\psi_{n3}, \phi_{n1})\phi_{n3} \otimes \phi_{n2}. \tag{4.92}$$

The annihilation operator thus acts like a derivation with alternating signs due to its fermionic nature. The relation of the two and three body states we have analyzed has a direct extension to the N-body case. The action of boson annihilation-creation operators can be derived in the same way.

Applying these operators to N and $N+1$ particle states, one finds directly ther commutation and anticommutation relations

$$[a(\psi_n), a^\dagger(\phi_n)]_\mp = (\psi_n, \phi_n), \tag{4.93}$$

where the $\mp$ sign, corresponds to commutator or anticommutator for the boson or fermion operators. If the functions ψ_n, ϕ_n belong to a normalized orthogonal set $\{\phi_{nj}\}$, then

$$[a(\phi_{ni}, a^\dagger(\phi_{nj}]_\mp = \delta_{ij}. \tag{4.94}$$

Let us now suppose that the functions ϕ_{nj} are plane waves in spacetime, *i.e.*, in terms of functions

$$\phi_{np}(x) = \frac{1}{(2\pi)^2}e^{-ip^\mu x_\mu}. \tag{4.95}$$

Then

$$(\phi_{np}, \phi_{np'}) = \delta^4(p - p'). \tag{4.96}$$

The quantum fields are then constructed as follows. Define

$$\phi_n(x) \equiv \int d^4p \, a(\phi_{np})e^{ip^\mu x_\mu}. \tag{4.97}$$

It then follows that, by the commutation (anticommutation) relations (3.52), these operators obey the relations

$$[\phi_n(x), \phi_n(x')]_{\mp} = \delta^4(x - x'), \qquad (4.98)$$

corresponding to the usual commutation relations of bose and fermion *fields*. Under Fourier transform, one finds the commutation relations in momentum space

$$[\phi_n(p), \phi_n(p')]_{\mp} = \delta^4(p - p') \qquad (4.99)$$

The relation of these quantized fields with those of the usual on-shell quantum field theories can be understood as follows. Let us suppose that the fourth component of the energy-momentum is $E = \sqrt{\mathbf{p}^2 + m^2}$, where m^2 is close to a given number, the on-shell mass of a particle. Then, noting that $dE = \frac{dm^2}{2E}$, if we multiply both sides of (4.97) by dE and integrate over the small neighborhood of m^2 occurring in both E and E', the delta function $\delta(E - E')$ integrates to unity. On the right hand side, there is a factor of $1/2E$, and we may absorb $\sqrt{dm^2}$ in each of the field variables, obtaining

$$[\phi_n(\mathbf{p}), \phi_n(\mathbf{p}')]_{\mp} = 2E\delta(\mathbf{p} - \mathbf{p}'), \qquad (4.100)$$

the usual formula for on-shell quantum fields. These algebraic results have been constructed in the foliation involved in the formulation of a consistent theory of relativistic spin, therefore admitting the action of the $SU(2)$ group for a many body system, applicable for unequal times.

It is clear from the construction of the Fock space that fields associated with different values of n^μ commute. The basis for the commutation relations is the creation and annihilation of (wave function) factors in the tensor product space; distinct values of n^μ therefore correspond to different species.

In the scalar product between states in the Fock space, one must complete the scalar products between functions by integrating over $\frac{d^3n}{n^0}$. A single value of n^μ in the product would have zero measure, so to compute probability amplitudes, one must construct wave packets over n^μ; these carry suitable weights for normalization. If the set $\{n\}$ is not a superselection rule, there would be transition matrix elements of observable connecting different values, and the form of the wave packets would play a physical role.

4.5 Discussion

In this chapter, we have discussed the concept of spin in the quantum theory and its manifestation in the framework of relativity. We have also shown how the properties observed in nature of symmetric (Bose–Einstein) and antisymmetric (Fermi–Dirac) states are associated with the particle spin. In this discussion, we have explained the method of induced representations of Wigner (1939).

In the next chapter, we discuss the two body bound state, also making essential use of the theory of induced repesentations.

Chapter 5

The Two Body Problem: Bound States

Models with action at a distance potentials, such as the Coulomb potential and the harmonic oscillator, have been very useful in nonrelativistic mechanics. They provide a simpler framework than the more fundamental field mediated models for interaction, and often provide a good approximation. They are also straightforwardly amenable to rigorous mathematical analysis.

The dynamical *phase space* of N particles contains the points $\mathbf{x}_n(t)$ and $\mathbf{p}_n(t)$, for $n = 1, 2, 3, \ldots N$; these points move through the phase space as a function of the parameter t, the Newtonian time, following some prescribed equations of motion. In this, what we might call a Newtonian-Galilean view, all events directly interact dynamically simultaneously.

For two particles interacting through a potential function $V(\mathbf{x}_1(t), \mathbf{x}_2(t))$; for Galilean invariance, for which the interactions do not change form under rotations or translations of the whole system, V would be a scalar function of the difference $\mathbf{x}_1(t) - \mathbf{x}_2(t)$. With the advent of special relativity, it became a challenge to formulate dynamical problems on the same level as that of the nonrelativistic theory.

For the relativistic theory, one might think of two world lines with action at a distance interaction, but the correlation that could be used between the two points x_1^μ and x_2^μ cannot be maintained by the variable t in every frame. Dirac (1932) introduced a "many time" theory to describe the dynamics of an N body system, maintaining the notion of the t component of the four vector position as associated with evolution, but the constraint theory [Horwitz (1982)] required to work in this framework poses difficult problems.

The Stueckelberg-Horwitz-Piron (SHP) theory, described in the previous chapters, provides an effective and systematic way of dealing with the N body problem, and has been applied in describing relativistic fluid mechanics [Sklarz (2001)], the Gibbs ensembles in statistical mechanics and the Boltzmann equation [Horwitz (1981)], systems of many identical particles, as described in Chapter 4, and other applications. The essential ingredient in developing these applications is the use of a single invariant parameter [Horwitz (1973)] to define the correlated interactions of a many body system.

We study in this chapter the relativistic quantum two body problem with invariant action at a distance potentials for bound states in the framework of the SHP theory. The two body quantum relativistic scattering problem provides an important and informative application for the methods developed in this chapter; we shall treat it in detail in the next chapter, devoted to general scattering theory.

5.1 The Two Body Bound State for Scalar Particles

For an invariant action at a distance potential for the two body relativistic bound state we assume the function $V(\rho)$, for

$$\rho^2 = (\mathbf{x}_1 - \mathbf{x}_2)^2 - (t_1 - t_2)^2 \equiv \mathbf{x}^2 - t^2, \qquad (5.1)$$

where x_1^μ and x_2^μ are taken at equal τ, acting as a correlation parameter as well as the global generating parameter of evolution. This "relative coordinate" (squared) reduces to $(\mathbf{x}_1 - \mathbf{x}_2)^2 \equiv \mathbf{x}^2$ at equal time for the two particles, in the nonrelativistic limit, so that $1/\rho$ becomes the Coulomb radial dependence $1/r$. Therefore, the solutions of a problem with this potential must then reduce to the solutions of the corresponding nonrelativistic problem in the non-relativistic limit, and therefore this problem poses an important challenge to the theory.

The two body Stueckelberg Hamiltonian is (see Chapter 2):

$$K = \frac{p_1{}^\mu p_{1\mu}}{2M_1} + \frac{p_2{}^\mu p_{2\mu}}{2M_2} + V(x). \qquad (5.2)$$

Since K does not depend on the total (spacetime) "center of mass",

$$X^\mu = \frac{M_1 x_1^\mu + M_2 x_2^\mu}{M_1 + M_2}, \qquad (5.3)$$

the two body Hamiltonian can be separated into the sum of two Hamiltonians, one for the "center of mass" motion and the other for the relative motion, by defining the total momentum, which is absolutely conserved,

$$P^\mu = p_1^\mu + p_2^\mu \qquad (5.4)$$

and the relative motion momentum

$$p^\mu = \frac{M_2 p_1^\mu - M_1 p_2^\mu}{M_1 + M_2}. \qquad (5.5)$$

This separation is *canonical*, *i.e.*, the pairs P^μ, X^μ and p^μ, x^μ separately satisfy the canonical Poisson bracket (classically) and commutation relations (quantum mechanically), and commute with each other.

Then, it is an identity that (as in the nonrelativistic two body problem)

$$K = \frac{P^\mu P_\mu}{2M} + \frac{p^\mu p_\mu}{2m} + V(x),$$

$$\equiv K_{CM} + K_{rel}, \qquad (5.6)$$

where $M = M_1 + M_2$, $m = M_1 M_2/(M_1 + M_2)$, and $x = x_1 - x_2$. Both K_{CM} and K_{rel} are constants of the motion. The total and relative momenta for the quantum case may be represented by partial derivatives with respect to the corresponding coordinates. This problem was solved explicitly for the classical case by Horwitz and Piron [Horwitz (1973)], where it was shown that there is no precession of the type predicted by Sommerfeld (1939) who used the nonrelativistic form $1/r$ for the potential (and obtained a period for the precession of Mercury that does not fit the data).

The corresponding quantum problem was solved by Cook (1972), with support for the wave functions in the full spacelike region; however, he obtained a spectrum that did not agree with the Balmer spectrum for hydrogen, *i.e.* a spectrum of the form $1/(n + \frac{1}{2})^2$, with

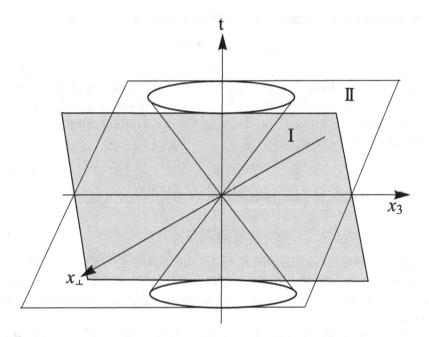

Fig. 2. Geometry of Reduced Minkowski Space (RMS), region external to planes I and II. (Courtesy Springer, Dordrecht).

n an integer. Zmuidzinas (1966), however, proved that there *is no complete set of functions* in the full spacelike region, and separated the spacelike region into two submanifolds, in each of which there could be complete orthogonal sets. This construction is shown in Fig. 2. The region for which $\mathbf{x}^2 > t^2$, in particular, permits the solution of the differential equations corresponding to the problem posed by (5.2) by separation of variables and provides spectra that coincide exactly with the corresponding nonrelativistic problems with potentials depending on r alone. We shall call this sector the RMS (reduced Minkowski space) [Arshansky (1989)]. We examine in the following the variables that span the full spacelike region, and after that, those that span the RMS.

The full spacelike region is spanned by

$$x^0 = \rho \sinh \beta, \quad x^1 = \rho \cosh \beta \cos \phi \sin \theta$$
$$x^2 = \rho \cosh \beta \sin \phi \sin \theta, \quad x^3 = \rho \cosh \beta \cos \theta \tag{5.7}$$

over all ρ from 0 to ∞, β in $(-\infty, \infty)$, ϕ in $(0, 2\pi)$ and θ in $(0, \pi)$. In general, separation of variables in a second order equation of d'Alembert type follows the method of separating out the least frequent variable on the right hand side of (5.7) first; the separation constant then enters into the second step, involving the separation of the next most frequent variable; this process is continued until all degrees of freedom have been accounted for. In the definitions given in (5.7), we see that the first separation constant, obtained from the equation for ϕ, which we could call m, enters into the equation for θ. The separation constant for the θ equation can be related to ℓ, as it occurs in the usual approach to the solution of the Laplace equation for the nonrelativistic case, and this constant then enters into the equation for β, a physical quantity not recognizable from the nonrelativistic theory (it is associated with hyperbolic functions involved in the Lorentz boost); the resulting quantum number, which we shall call n, becomes involved in the equation involving the "radial" coordinate. The solution of the equation for the radial function provides the spectrum, in this case, with dependence on a variable not easily associated with the nonrelativistic (and experimentally satisfactory) results.

On the other hand, the set of variables describing the RMS, running over the same range of parameters,

$$x^0 = \rho \sin\theta \sinh\beta, \quad x^1 = \rho \sin\theta \cosh\beta \cos\phi$$
$$x^2 = \rho \sin\theta \cosh\beta \sin\phi, \quad x^3 = \rho \cos\theta, \tag{5.8}$$

cover the entire space external to the planes shown in Fig. 2 (for $x_1^2 + x_2^2 > t^2$).

As for (5.7), for $\beta \to 0$, these coordinates become the standard spherical representation of the three dimensional space (at the "simultaneity" point $t = 0$, where ρ becomes r). However, the sequence of separation of variables corresponds to the assignment of the constant m, then n, and finally ℓ, which is, in this case, the only separation constant entering the spectrum determining equation for the ρ dependence in the solution. This constant, associated with the Legendre functions, then carries the physical meaning of angular momentum, entering in the same way in the spectrum for the non-relavistic case, as we shall see. Independently of the form of the potential $V(\rho)$, one obtains the same radial equation (in ρ) as for the nonrelativistic Schrödinger equation (in r), and therefore the same spectra for the reduced Hamiltonian. We shall discuss the relation of these results to the energy spectrum after writing the solutions. We summarize in the following the basic mathematical steps.

Assuming the total wavefunction (for $P \to P'$, a point on the continuum of the spectrum of the conserved operator P)

$$\Psi_{P'\tau}(X,x) = e^{iP'^\mu X_\mu}\psi_{P'\tau}(x), \qquad (5.9)$$

the evolution equation for each value of the total energy momentum of the system then becomes

$$i\frac{\partial}{\partial\tau}\Psi_{P'\tau}(X,x) = (K_{CM} + K_{rel})\Psi_{P'\tau}(X,x)$$

$$= \left[\frac{P'^2}{2M} + K_{rel}\right]\Psi_{P'\tau}(X,x). \qquad (5.10)$$

For the case of discrete eigenvalues K_a of K_{rel}, due to the complete separation of variables into total and relative motion, the discrete eigenfunctions do not depend explicitly on P'. The Hilbert space of the reduced motion is attached to the point P' on the spectrum of K_{CM}), we then have (we suppress reference to the value of P' in the following) the eigenvalue equation (cancelling the center of mass wave function factor and K_{CM} on both sides)

$$K_{rel}\psi^{(a)}(x) = K_a\psi^{(a)}(x)$$

$$= (-(1/2m)\partial_\mu\partial^\mu + V(\rho))\psi^{(a)}(x), \qquad (5.11)$$

resembling a Klein–Gordon type equation with a spacetime dependent "mass" term.

Since K commutes with the generators $M_{\mu\nu}$, we can construct two quadratic scalar operators, called *Casimir operators* that are constants of the motion and serve to label the states of the system.

In general, continuous groups are characterized by quadratic forms that are left invariant under the group action. The square of the total angular momentum operator is the Casimir operator of the rotation group $O(3)$ in the nonrelativistic case.

There are two Casimir operators defining the representations of $O(3,1)$ [Naimark (1964)]. The first Casimir operator is

$$\Lambda = \frac{1}{2} M_{\mu\nu} M^{\mu\nu}; \tag{5.12}$$

the second Casimir operator $\frac{1}{2} \epsilon^{\mu\nu\lambda\sigma} M_{\mu\nu} M_{\lambda\sigma}$ is identically zero for two particles without spin. Recalling that our separation into center of mass and relative motion is canonical, and that

$$M^{\mu\nu} = x^{\mu} p^{\nu} - x^{\nu} p^{\mu}, \tag{5.13}$$

it is straighforward to show, using the canonical commutation relations, that

$$\Lambda = x^2 p^2 + 2ix \cdot p - (x \cdot p)^2. \tag{5.14}$$

Since

$$x \cdot p \equiv x^{\mu} p_{\mu} = -i\rho \frac{\partial}{\partial \rho}, \tag{5.15}$$

we see that

$$\Lambda = -\rho^2 \partial^{\mu} \partial_{\mu} + 3\rho \frac{\partial}{\partial \rho} + \rho^2 \frac{\partial^2}{\partial \rho^2},$$

or

$$-\partial_{\mu} \partial^{\mu} = -\frac{\partial^2}{\partial \rho^2} - \frac{3}{\rho} \frac{\partial}{\partial \rho} + \frac{\Lambda}{\rho^2}. \tag{5.16}$$

It now follows that (5.11) can be written as

$$K_a \psi^{(a)}(x) = \left\{ \frac{1}{2m} \left[-\frac{\partial^2}{\partial \rho^2} - \frac{3}{\rho} \frac{\partial}{\partial \rho} + \frac{\Lambda}{\rho^2} \right] + V(\rho) \right\} \psi^{(a)}(x). \tag{5.17}$$

Since

$$L_i = \frac{1}{2} \epsilon_{ijk} (x^j p^k - x^k p^j), \tag{5.18}$$

corresponding to the definition of the nonrelativistic angular momentum $\mathbf{L}$, and

$$A^i = x^0 p^i - x^i p^0, \tag{5.19}$$

corresponding to the boost generator $\mathbf{A}$, we obtain

$$\Lambda = \mathbf{L}^2 - \mathbf{A}^2. \tag{5.20}$$

One then finds, after writing the time and spatial derivatives in terms of definitions that we have given in (5.8),

$$\Lambda = -\frac{\partial^2}{\partial\theta^2} - 2\cot\theta\frac{\partial}{\partial\theta} + \frac{1}{\sin^2\theta}N^2, \tag{5.21}$$

where

$$N^2 = L_3^2 - A_1^2 - A_2^2 \tag{5.22}$$

is the Casimir operator of the $O(2,1)$ subgroup of $O(3,1)$ leaving the z axis (and the RMS submanifold) invariant [Bargmann (1947)]. In terms of the RMS variables that we have defined in (5.8),

$$N^2 = \frac{\partial^2}{\partial\beta^2} + 2\tanh\beta\frac{\partial}{\partial\beta} - \frac{1}{\cosh^2\beta}\frac{\partial^2}{\partial\phi^2}. \tag{5.23}$$

We emphasize that these operators act freely on their complete natural domain on the whole range of the coordinate parametrizations. Except for the derivatives on β, which runs on the whole real line, derivative with respect to ρ, θ and ϕ have the same problems of Hermiticity as for the usual nonrelativistic spherical coordinates (these variables are bounded or semibounded, and the derivatives are not *a priori* defined at the end points). However, as in the nonrelativistic case, the second order operators we shall use to characterize the quantum states are essentially self-adjoint, and one obtains real spectra (in general, non-self adjoint operators may have complex sprctra).

We now proceed to separate variables and find the eigenfunctions. The solution of the general eigenvalue problem (5.17) can be written

$$\psi(x) = R(\rho)\Theta(\theta)B(\beta)\Phi(\phi), \tag{5.25}$$

with invariant measure in the $L^2(R^4)$ of the RMS

$$d\mu = \rho^3\sin^2\theta\cosh\beta d\rho d\phi d\beta d\theta. \tag{5.26}$$

The functions which are factors of the separated solution (5.25) must each be normalized on its range. To satisfy the ϕ derivatives in (5.23), it is necessary to take

$$\Phi_m(\phi) = \frac{1}{\sqrt{2\pi}} e^{i[m+\frac{1}{2}]\phi}, \quad 0 \le \phi < 2\pi, \tag{5.27}$$

where we have indexed the solutions by the separation constant m. For the case m an integer, this is a *double valued function*. To be compatible with the conditions on the other functions, this is the necessary choice; one must use $\Phi_m(\phi)$ for $m \ge 0$ and $\Phi_m^*(\phi)$ for $m < 0$.

It was suggested by Bacry (1990) that the occurrence of the half-integer in the phase can be understood in terms of the fact that the RMS is a connected, but not simply connected manifold. One can see this by considering the projective form of the restrictions

$$x^2 + y^2 + z^2 - t^2 > 0 \tag{5.28}$$

assuring that the events are relatively spacelike, and

$$x^2 + y^2 - t^2 > 0, \tag{5.29}$$

assuring, in addition, that the relative coordinates lie in the RMS. Dividing (5.28) and (5.29) by t^2, and calling the corresponding projective variables X, Y, Z, we have from (5.28)

$$X^2 + Y^2 + Z^2 > 1, \tag{5.30}$$

the exterior of the unit sphere in the projective space, and from (5.29),

$$X^2 + Y^2 > 1, \tag{5.31}$$

the exterior of the unit cylinder along the z-axis. Since the space is projective, we can identify the points of the cylinder at infinity, and see that this corresponds to a torus with the unit sphere around the origin imbedded in the torus. Thus, a closed curve around the torus, passing through the central region, cannot be homotopically (by continuous deformation) contracted to a point, indicating the region is not simply connected. Characteristically, such a topological structure is associated with half integer phase (*e.g.* [Shapere (1989)]). This picture also provides a simple interpretation of what would happen to a quantum state with wave packet inside the torus (in the region $X^2 + Y^2 < 1$); it could tunnel through the imbedded sphere, continuously connected to this interior solution, which would then be a scattering state, not a bound state.

We now continue with our discussion of the structure of the solutions.

The operator Λ contains the $O(2,1)$ Casimir N^2; with our solution (5.23), we then have

$$
N^2 B_{mn}(\beta) = \left[\frac{\partial^2}{\partial \beta^2} + 2\tanh\beta \frac{\partial}{\partial \beta} + \frac{(m + \frac{1}{2})^2}{\cosh^2 \beta} \right] B_{mn}(\beta)
$$

$$
\equiv \left(n^2 - \frac{1}{4} \right) B_{mn}(\beta), \tag{5.32}
$$

where n^2 is the separation constant for the variable β. The term $(m + \frac{1}{2})^2$ must be replaced by $(m - \frac{1}{2})^2 = (|m| + (\frac{1}{2}))^2$ for $m < 0$. We study only the case $m \geq 0$ in what follows. The remaining equation for Λ is then

$$
\Lambda\Theta(\theta) = \left[-\frac{\partial^2}{\partial \theta^2} - 2\cot\theta \frac{\partial}{\partial \theta} + \frac{1}{\sin^2 \theta} \left(n^2 - \frac{1}{4} \right) \right] \Theta(\theta). \tag{5.33}
$$

For the treatment of Eq. (5.32), it is convenient to make the substitution

$$
\zeta = \tanh\beta, \tag{5.34}
$$

so that $-1 \leq \zeta \leq 1$. One then finds that for

$$
B_{mn}(\beta) = (1 - \zeta^2)^{1/4} \hat{B}_{mn}(\zeta), \tag{5.35}
$$

(5.32) becomes the well-known equation

$$
(1 - \zeta^2) \frac{\partial^2 \hat{B}_{mn}(\zeta)}{\partial \zeta^2} - 2\zeta \frac{\partial \hat{B}_{mn}(\zeta)}{\partial \zeta}
$$

$$
+ \left[m(m + 1) - \frac{n^2}{1 - \zeta^2} \right] \hat{B}_{mn}(\zeta) = 0. \tag{5.36}
$$

The solutions are the associated Legendre functions of the first and second kind [Gel'fand (1963)][Merzbacher (1970)], $P_m^n(\zeta)$ and $Q_m^n(\zeta)$.

The normalization condition on these solutions, with the measure (5.26) is

$$\int \cosh \beta |B(\beta)|^2 d\beta < \infty,$$

or, in terms of the variable ζ

$$\int_{-1}^{1} (1 - \zeta^2)^{-1} |\hat{B}(\zeta)|^2 d\zeta < \infty. \tag{5.37}$$

The second kind Legendre functions do not satisfy this condition.

For the condition on the $P_m^n(\zeta)$, it is simplest to write the known result [Gradstyn (2018)]

$$\int_{-1}^{1} (1 - \zeta^2)^{-1} |P_{\mu+\nu}^{-\nu}(\zeta)|^2 d\zeta = \frac{1}{\nu} \frac{\Gamma(1 + \mu)}{\Gamma(1 + \mu + 2\nu)} \tag{5.38}$$

The normalized solutions (it is sufficient to consider $n \geq 0$) may be written as

$$\hat{B}_{mn}(\zeta) = \sqrt{n} \sqrt{[\Gamma(1 + m + n)/\Gamma(1 + m - n)]} \times P_m^{-n}(\zeta), \tag{5.39}$$

where $m \geq n$.

It is very significant for the structure of the theory that the case $n = 0$ must be treated with special care; it requires a regularization. For $n = 0$, the associated Legendre functions become the Legendre polynomials $P_m(\zeta)$. In terms of the integration on β, the factor $\cosh \beta = (1 - \zeta^2)^{-1/2}$ (so that $d\beta = \frac{1}{(1-\zeta^2)^{\frac{3}{2}}} d\zeta$) in the measure is cancelled by the square of the factor $(1 - \zeta^2)^{1/4}$ in the norm; the integration then appears as

$$\int_{-\infty}^{\infty} |\hat{B}_m(\zeta)|^2 d\beta.$$

   The Legendre polynomials do not vanish at $\zeta = \pm 1$, so if $\hat{B}_m$ and P_m are related by a finite coefficient, the integral would diverge. When n goes to zero, as we shall see, associated with the ground state, the wave function spreads along the hyperbola labelled by ρ, going asymptotically to the light plane; the probability density with respect to intervals of β becomes constant for large β. Events associated with the two particles may be found (for sufficiently large space separation) with 2+1 lightlike separation out to remote regions of the tangent planes. There is no four-current associated with such a bound state, and therefore one would not see individual particles. The configuration provides a spacetime structure to the ground state. The extended configuration of the wave function requires "regularization" to obtain a finite norm. This an important and fundamental property of the relativistic theory, with strong consequences for the structure of the representaions of the Lorentz group that the bound states provide.

The (regularized) expectation values reproduce the distribution of the lowest Schrödinger bound state, although the spacetime wave function approaches that of a generalized eigenfunction in the time direction.

To carry out the regularization, we take the limit as n goes continuously to zero after computation of scalar products. Thus, we assume the form

$$\hat{B}_m(\zeta) = \sqrt{\epsilon}(1 - \zeta^2)^{\epsilon/2} P_m(\zeta), \qquad (5.40)$$

with $\epsilon \to 0$ after computation of scalar products. This formula is essentially a residue of the Rodrigues formula

$$P_m^{-n}(\zeta) = (-1)^n (1 - \zeta^2)^{n/2} \frac{d^n}{d\zeta^n} P_m(\zeta) \qquad (5.41)$$

for $n \to 0$.

These remarks are, as indicated above, important for the structure of the theory. The operator for the differential equation (5.17) for the eigenvalue of the reduced motion is invariant under the action of the Lorentz group. It follows from acting on the equation with the unitary representation of the Lorentz group that the eigenfunctions must be representations of that group [Wigner (1931)] for each value of the eigenvalue. However, as one can easily see, the solutions that we found are, in fact, irreducible representations of $O(2,1)$ (due to the fact that the support manifold is in a $2 + 1$ space, x, y, t), not, *a priori*, representations of the Lorentz group $O(3,1)$.

To classify the states, we have required that the wave functions be eigenfunctions of the Casimir operator (5.22) of the $O(2,1)$ subgroup.

For the generators of $O(2,1)$, we note that

$$H_\pm \equiv A_1 \pm iA_2 = e^{\pm i\phi}\left(-i\frac{\partial}{\partial\beta} \pm \tanh\beta\frac{\partial}{\partial\phi}\right),$$

$$L_3 = -i\frac{\partial}{\partial\phi},$$

$$A_3 = -i\left(\cot\theta\cosh\beta\frac{\partial}{\partial\beta} - \sinh\beta\frac{\partial}{\partial\theta}\right),$$

$$L_\pm = L_1 \pm iL_2$$

$$= e^{\pm i\phi}\left(\pm\cosh\beta\frac{\partial}{\partial\theta} - \sinh\beta\cot\theta\frac{\partial}{\partial\beta}\right.$$

$$\left. + i\frac{\cot\theta}{\cosh\beta}\frac{\partial}{\partial\phi}\right). \tag{5.42}$$

It then follows that $H_\pm$ are raising and lowering operators for m on the functions

$$\xi_{n+k}^{-n}(\zeta,\phi) \equiv B_{n+k,n}(\beta)\Phi_{n+k}(\phi)$$

$$= (1-\zeta^2)^{1/4}\hat{B}_{n+k,n}(\zeta)\Phi_{n+l}(\phi), \tag{5.43}$$

where it is convenient to replace m by $n+k$. With the relation

$$[L_3, H_\pm] = \pm H_\pm \tag{5.44}$$

one can show [Arshansky (1989)] that

$$H_+\chi_{n+k}^{-n}(\zeta,\phi) = i\sqrt{(k+1)(2n+k+1)}\chi_{n+k+1}^{-n}(\zeta,\phi) \tag{5.45}$$

and that

$$H_-\chi_{n+k+1}^{-n}(\zeta,\phi) = -i\sqrt{(k+1)(2n+k+1)}\chi_{n+k}^{-n}(\zeta,\phi). \tag{5.46}$$

The complex conjugate of χ_{n+k}^{-n} transforms in a similar way, resulting in a second (inequivalent) representation of $O(2,1)$ with the same value of the $O(2,1)$ Casimir operator (these states correspond to replacement of $m+\frac{1}{2}$ by $m-\frac{1}{2}$ for $m < 0$, and are the result of "charge conjugation") as we show below. Since the operators A_1, A_2 and L_3 are Hermitian, complex conjugation is equivalent to the transpose. Replacing these operators by their negative transpose (to be defined

by C), leaves the commutation relations invariant. Thus the action
on the complex conjugate states involves

$$H_-^C = -H_+^* = H_-, \quad H_+^C = -H_-^* = H_+,$$
$$L_3^C = -L_3^* = L_3; \tag{5.47}$$

These are precisely the operators under which the complex conjugate states transform, and thus corresponds to "charge conjugation"(defined as taking all additive quantum numbers to their
negative).

We have therefore determined that the wave functions we have
obtained are irreducible representations of $O(2,1)$. To construct
representations of $O(3,1)$, let us consider the well established method
which is effective in constructing representations of $O(3,1)$ from
representations of $O(3)$, a group that we would have found if
we were working with solutions in the timelike region [Naimark
(1964), Gel'fand (1963)], called the *ladder representation*. It follows
from the Lie algebra of $O(3,1)$ that the $O(3)$ subgroup Casimir
operators $\ell(\ell+1)$ are stepped by $\ell \to \ell \pm 1$ under the action of the
boost from $O(3,1)$. Thus the whole set of representations of $O(3)$,
from $\ell = 0$ to ∞ form a representation of $O(3,1)$. Each of the representations of $O(3)$ entering this tower are trivially normalizable, since
they are of dimension $(2\ell + 1)$. However, attempting to apply this
method to the representations of $O(2,1)$ fails, since the normalization
of these representations is far from trivial; they are represented in
infinite dimensional Hilbert spaces, since there are no unitary finite
dimensional representations of a noncompact group such as $O(2,1)$.
The application of the Lie algebra to this set connects the lowest
state of the tower with the ground state which, as we have shown,
requires regularization. The action of the algebra does not provide
such a regularization, and therefore the method is inapplicable.

We have discussed in Chapter 4 the idea of the *induced representation*, there, applied to representations of spin based on a timelike
vector. We may apply this method to contructing the representations
of $O(3,1)$ based on an induced representation with the $O(2,1)$ "little
group", based on a spacelike vector corresponding to the choice of
the z axis. We shall follow this method in the next section.

We now return to record the solutions of (5.33).

Defining

$$\xi = \cos\theta \tag{5.48}$$

and the functions

$$\hat{\Theta}(\theta) = (1 - \xi^2)^{1/4}\Theta(\theta), \tag{5.49}$$

Eq. (5.33) becomes

$$\frac{d}{d\xi}\left((1-\xi^2)\frac{d}{d\xi}\hat{\Theta}(\theta)\right) + \left(\ell(\ell+1) - \frac{n^2}{1-\xi^2}\right)\hat{\Theta}(\theta), \tag{5.50}$$

where we have defined

$$\Lambda = \ell(\ell+1) - \frac{1}{4}. \tag{5.51}$$

The solutions of (5.50) are proportional to the associated Legendre functions of the first or second kind, $P_\ell^n(\xi)$ or $Q_\ell^n(\xi)$. For $n \neq 0$, the functions of the second kind, as pointed out above, are not normalizable (with the measure (5.26)) and we therefore reject these.

The unitary irreducible representations of $O(2,1)$ are single or double valued, and hence m must be integer or half integer. As we have seen (see (5.3)), k is integer valued, and therefore n must be integer or half integer also. Normalizability conditions on the associated Legendre functions then require that ℓ be respectively, positive half-integer or integer. The lowest mass state, as we shall see from the spectral results, corresponds to $\ell = 0$, and hence we shall consider only integer values of ℓ. Note that in the nonrelativistic quantum theory the angular momentum quantum number ℓ is chosen to be integer to provide correct representations for the rotation group [Gottfried (1962)]. Therefore, n and m must be integer.

If we take β to zero in the RMS variables of (5.8), this set of variables, as we pointed out, reduces to the standard spherical coordinates on $3D$. The factor

$$Y_\ell^n(\theta, \phi) = \frac{1}{\sqrt{2\pi}}e^{i\phi n}\hat{\Theta}_\ell^n(\theta) \tag{5.52}$$

in the separated solution, where

$$\hat{\Theta}_\ell^n(\theta) = \left(\frac{2\ell+1}{2}\frac{(\ell-n)!}{(\ell+n)!}\right)^{1/2} P_\ell^n(\cos\theta) \tag{5.53}$$

transforms as

$$Y_\ell^n(\theta, \phi) = \Sigma^{n'} D_{mn'}^\ell(\eta_1, \eta_2, \eta_3) Y_\ell^{n'}(\theta', \phi'), \qquad (5.54)$$

where the $D_{mn'}^\ell$ are the Wigner rotation functions [Wigner (1931)] of the Euler angles η_1, η_2, η_3.

We now turn to the solution of the radial equations, containing the spectral content of the theory. With the evaluation of Λ in (5.51), we may write the radial equation as

$$\left[\frac{1}{2m}\left(-\frac{\partial^2}{\partial\rho^2} - \frac{3}{\rho}\frac{\partial}{\partial\rho} + \frac{\ell(\ell+1) - \frac{3}{4}}{\rho^2}\right) + V(\rho)\right] R^{(a)}(\rho)$$

$$= K_a R^{(a)}(\rho). \qquad (5.55)$$

If we put

$$R^{(a)}(\rho) = \frac{1}{\sqrt{\rho}}\hat{R}^{(a)}(\rho), \qquad (5.56)$$

Eq. (5.55) becomes precisely the nonrelativistic Schrödinger equation for $\hat{R}^{(a)}$ in the variable ρ, with potential $V(\rho)$ (the measure for the norm in the Hilbert space for these functions is, from (5.26), just $\rho^2 d\rho$, as for the nonrelativistic theory)

$$\frac{d^2\hat{R}^{(a)}(\rho)}{d\rho^2} + \frac{2}{\rho}\frac{d\hat{R}^{(a)}(\rho)}{d\rho} - \frac{\ell(\ell+1)}{\rho^2}\hat{R}^{(a)}(\rho)$$

$$+ 2m(K_a - V(\rho))\hat{R}^{(a)}(\rho) = 0. \qquad (5.57)$$

The lowest eigenvalue K_a. as for the energy in the nonrelativistic Schrödinger equation, corresponds to the $\ell = 0$ state of the sequence $\ell = 0, 1, 2, 3, \ldots$, and therefore the quantum number ℓ plays a role analogous to the orbital angular momentum. This energy is of a lower value than achievable with wave functions with support in the full spacelike region [Cook (1972)] and the relaxation of the system to wave functions with support in the RMS may be thought of, in this sense, as a spontaneous symmetry breaking (I thank A. Ashtekar for his remark on this point [Ashtekar (1982)]).

We emphasize that this type solution is available for *every nonrelativistic problem with spherically symmetry potential* $V(r)$; all of the details of our derivation depend only on the angular and hyperangular properties of the Stueckelberg–Schrödinger operator.

The value of the full generator K is then determined by these eigenvalues and the value of the center of mass total mass squared operator, *i.e.*,

$$K = \frac{P^\mu P_\mu}{2M} + K_a. \tag{5.58}$$

The first term corresponds to the total effective rest mass of the system, and contains the observable energy spectrum through the mass energy relation of Einstein. In particular, the invariant mass squared of the system is given by (sometimes called the Mandelstam variable s [Chew (1966)])

$$s_a \equiv -P_a^2 = 2M(K_a - K). \tag{5.59}$$

This total center of mass momentum is observed in the laboratory in scattering and decay processes, where it is defined as the sum of the outgoing momenta squared (see Chapter 6). In the case of two particles, it would be given by $-(p_1^\mu + p_2^\mu)(p_{1\mu} + p_{2\mu})$, as we have defined it in (5.59). In terms of total energy and momentum,

$$s_a = E_T^2 - \mathbf{P}_T^2, \tag{5.60}$$

and in the center of momentum frame, for $\mathbf{P} = 0$, is just E_T^2.

In order to extract information about the *energy spectrum*, we must therefore make some assumption on the value of the conserved quantity K. In the case of a potential that vanishes for large ρ, we may consider the two particles to be asymptotically free, so the effective Hamiltonian in this asymptotic region

$$K \cong \frac{p_1{}^\mu p_{1\mu}}{2M_1} + \frac{p_2{}^\mu p_{2\mu}}{2M_2}. \tag{5.61}$$

Further, assuming that the two particles at very large distances, in accordance with our experience, undergo a relaxation to their mass shells, so that $p_i^2 \cong -M_i^2$ (although the mechanism for this effect is not definitively known, the radiation reaction problem [Dirac (1947)] and some results in statistical mechanics provide examples of how such a mechanism could work). In this case, K would be assigned the value

$$K \cong -\frac{M_1}{2} - \frac{M_2}{2} = -\frac{M}{2}.$$

The two particles in this asympotic state would, for the bound state problem, be at the ionization point. The process of bringing the two particles together by some interaction localized remotely far in a timelike direction, say, for large negative values of τ, which would not influence the solutions of the bound state problem appreciably, could then adiabatically bring the particles into their bound states without affecting the conserved total K. If these assumptions are approximately valid, we find for the total energy, which we now label E_a,

$$E_a/c \cong \sqrt{M^2c^2 + 2MK_a}, \tag{5.62}$$

where we have restored the factors c.

In the case of excitations small compared to the total mass of the system, we may factor out Mc and represent the result as a power series expansion

$$E_a \cong Mc^2 + K_a - \frac{1}{2}\frac{K_a}{Mc^2} + \cdots, \tag{5.63}$$

so that the energy spectrum is just the set $\{K_a\}$ up to relativistic corrections. Thus, the spectrum for the $1/\rho$ potential is just that of the nonrelativistic hydrogen problem up to relativistic corrections, of order $1/c^2$.

If the spectral set $\{K_a\}$ includes large negative values, the result (5.62) could become imaginary, indicating the possible onset of instability. The asymptotic condition imposed on the evaluation of K must be re-examined in this case. If the potential grows very rapidly as $\rho \to 0$, then at large spacelike distances, where the hyperbolic surfaces $\rho = const$ approach the lightcone, the Euclidean measure d^4x (thought of, in this context, as small but finite) on the R^4 of spacetime covers very singular values and the expectation values of the Hamiltonian at large spacelike distances may not permit the contribution of the potential to become negligible; it may have an effectively very long range. This effect can occur in the transverse direction to the z axis along the tangent to the light cone; the hyperbolas cannot reach the light cone in the z direction, which, as we shall see in Chapter 6, corresponds to the direction of a scattering beam in the standard phase shift analysis. It may play an important role in the modelling the behavior of the transverse scattering amplitudes in high energy scattering studied, for example, by Hagedorn (1965).

The resulting value of K, perhaps at this level necessarily chosen phenomenologically to fit the data, may therefore nevertheless maintain real values for E_a. This question constitutes an interesting field for rigorous analysis [thanks S. Nussinov for the discussion], somewhat related to the existence of the wave operator in scattering theory, which we will discuss in Chapter 6.

5.2 Some Examples

In this section we give some examples. We will treat the Coulomb potential, the oscillator and the analog of the three dimensional square well.

For the analog of the Coulomb potential, we take

$$V(\rho) = -\frac{Ze^2}{\rho}. \tag{5.64}$$

In this case the spectrum, according to the solutions above, is given by

$$K_a = -\frac{Z^2 me^4}{2\hbar^2(\ell + 1 + n_a)^2}, \tag{5.65}$$

where $n_a = 0, 1, 2, 3 \ldots$. The wave functions $\hat{R}(\rho)^a$ are the usual hydrogen functions (*e.g.* [Landau (1965)])

$$\hat{R}_{n_a\ell}(\rho) = \sqrt{\frac{Zn_a!}{(n_a + \ell + 1)^2(n_a + 2\ell + 1)}} e^{-x/2} x^{\ell+1} L_{n_a}^{2\ell+1}(x), \tag{5.66}$$

where $L_{n_a}^{2\ell+1}$ are the Laguerre polynomials, and the variable x is defined by

$$x = \frac{(2Z\rho/a_0)}{(n_a + \ell + 1)}, \tag{5.67}$$

and $a_0 = \hbar^2/me^2)$. The *size* of the bound state, whic is related to the atomic form factor, is measured according to the variable ρ [Hofstadter (1958)]. For the lowest level (using the regularized functions) $n_a = \ell = 0$,

$$\langle\rho\rangle_{n_a=\ell=0} = \frac{3}{2}a_0. \tag{5.68}$$

The total mass spectrum, given by (5.59), is then

$$s_{n_a,\ell} \cong M^2 c^2 - \frac{mMZ^2 e^4}{\hbar^2(n_a + \ell + 1)^2}. \tag{5.69}$$

For the case that the nonrelativistic spectrum has values small compared to the sum of the particle rest masses, we may use the approximate relation (5.63) to obtain

$$E_{a,\ell} \cong Mc^2 - \frac{Z^2 me^4}{2\hbar^2(n_a + \ell + 1)^2}$$

$$- \frac{1}{8}\frac{Z^4 m^2 e^8}{Mc^2\hbar^4(n_a + \ell + 1)^4} + \cdots. \tag{5.70}$$

The lowest order relativistic correction to the rest energy of the two body system with Coulomb like potential is then

$$\frac{\Delta(E_{a,\ell} - Mc^2)}{E_{a,\ell} - Mc^2} = \frac{Z\alpha^2}{4}\left(\frac{m}{M}\right)\frac{1}{(n_a + \ell + 1)^2}. \tag{5.71}$$

For spinless atomic hydrogen ($Z = 1$), $\Delta(E - Mc^2) \cong 9.7 \times 10^{-8}\,\mathrm{eV}$ and $E - Mc^2 \cong 13.6\,\mathrm{eV}$ for the ground state. The relativistic correction is therefore of the order of one part in 10^8, about 10% of the hyperfine splitting. For positronium, $\Delta(E - Mc^2) \sim 2 \times 10^{-5}\,\mathrm{eV}$ it is about one part in 10^5, about 2% of the positronium hyperfine splitting of $8.4 \times 10^{-4}\,\mathrm{eV}$ [Itzykson (1980)]. We see quantitatively that the relativistic theory gives results that are consistent with the known data on these experimentally well studied bound state systems.

For the four dimensional oscillator, with $V(\rho) = \frac{1}{2}m\omega^2\rho^2$, Eq. (5.57) takes the form

$$\frac{d^2\hat{R}^{(a)}(\rho)}{d\rho^2} + \frac{2}{\rho}\frac{d\hat{R}^{(a)}(\rho)}{d\rho} - \frac{\ell(\ell+1)}{\rho^2}\hat{R}^{(a)}(\rho)$$

$$+ 2m\left(K_a - \frac{m^2\omega^2}{\hbar^2}\rho^2 - \frac{\ell(\ell+1)}{\rho^2}\right)\hat{R}^{(a)}(\rho) = 0. \qquad (5.72)$$

With the transformation

$$\hat{R}^{(a)}(\rho) = x^{\ell/2}e^{-x/2}w^{(a)}(x), \qquad (5.73)$$

for

$$x = \frac{m\omega}{\hbar}\rho^2, \qquad (5.74)$$

we obtain the equation

$$x\frac{d^2w^{(a)}}{dx^2} + \left(\ell + \frac{3}{2} - x\right)\frac{dw^{(a)}}{dx}$$

$$+ \frac{1}{2}\left(\ell + \frac{3}{2} - \frac{K_a}{\hbar\omega}\right)w^{(a)} = 0 \qquad (5.75)$$

Normalizable solutions, the Laguerre polynomials $L_{n_a}^{\ell+1/2}(x)$, exist [Landau (1965)] when the coefficient of $w^{(a)}(x)$ is a negative integer, so that the eignvalues are

$$K_a = \hbar\omega\left(\ell + 2n_a + \frac{3}{2}\right), \qquad (5.76)$$

where $n_a = 0, 1, 2, 3, \ldots$ The total mass spectrum is given by (5.59) as

$$s_{n_a,\ell} = -2MK + 2M\hbar\omega\left(\ell + 2n_a + \frac{3}{2}\right), \qquad (5.76)$$

Note that the "zero point" term is $\frac{3}{2}$, indicating that in the RMS, in the covariant equations there are effectively three intrinsic degrees of freedom, as for the nonrelativistic oscillator.

The choice of K is arbitrary here, since there is no ionization point for the oscillator, and no *a priori* way of assigning it a value; setting $K = -\frac{Mc^2}{2}$ as for the Coulomb problem (a choice that may be justified by setting the spring constant equal to zero and adiabatically increasing it to its final value), one obtains, for small excitations relative to the particle masses,

$$E_a \cong Mc^2 + \hbar\omega \left(\ell + 2n_a + \frac{3}{2} \right)$$

$$-\frac{1}{2} \frac{\hbar^2 \omega^2 \left(\ell + 2n_a + \frac{3}{2} \right)^2}{Mc^2} + \cdots \qquad (5.77)$$

Feynman, Kislinger and Ravndal [Feynman (1971)], Kim and Noz [Kim (1977)] and Leutwyler and Stern [Leutwyler (1977)] have studied the relativistic oscillator and obtained a positive spectrum (as in (5.76)) by imposing a subsidiary condition suppressing timelike excitations, which lead (in the formalism of annihilation-creation operators to generate the spectrum) to negative norm states ("ghosts"). There are no ghost states in the covariant. SHP treatment, and no extra constraints invoked in finding the spectrum. The solutions are given in terms of Laguerre poynomials, but unlike the case of the standard treatment of the $4D$ oscillator, in which $x^\mu \pm ip^\mu$ are considered annihilation-creation operators, the spectrum generating algebra (for example, Dothan (1965)) for the covariant SHP oscillator has been elusive [Land (2011)].

We now turn to the $O(3,1)$ invariant square well. In this case the radial equation (5.57) must be solved with the potential given by:

$$V(\rho) = -U, \quad \rho \le a \qquad (5.78)$$

and

$$V(\rho) = 0, \quad \rho > a \qquad (5.79)$$

The solutions have the form (for $-U \le K_a \le 0$) [Merzbacher (1970)]:

$$\hat{R}^{(a)}(\rho) = Aj_\ell(\kappa_1 \rho), \quad \rho \le a, \qquad (5.80)$$

and

$$\hat{R}^{(a)}(\rho) = Bh_\ell^{(1)}(i\kappa_2 \rho), \quad \rho > a \qquad (5.81)$$

where j_ℓ are spherical Bessel functions and $h_\ell^{(1)}$ are spherical Hankel functions of the first kind; here

$$\kappa_1 = \sqrt{2m(K_a + U)/\hbar^2}, \quad \kappa_2 = \sqrt{(-2mK_a)/\hbar^2} \qquad (5.82)$$

The radial measure for $\hat{R}^{(a)}(\rho)$ is the same as for the nonrelativistic case (in r), *i.e.*, $\rho^2 d\rho$. Continuity of the wave function at the boundary $\rho = a$ provides the condition for the values of K_a. Note that the boundary at $\rho = a$ for the square well is the apex of the hyperbola; at $t = 0$, it lies at $r = a$. For the arguments of the Bessel functions large enough (high excitations, a sufficiently large) the asymptotic form of the Bessel functions may be used to investigate some analytic properties of the wave functions. Since $\kappa_1^2 + \kappa_2^2 = 2mU/\hbar^2$, the large argument approximation is satisfied for

$$\xi^2 \equiv \frac{2mU}{\hbar^2} a^2 \gg 1. \tag{5.83}$$

Using the asymptotic forms

$$j_\ell(z) \sim \frac{1}{z}\cos(z - \pi\ell/2 - \pi/2)$$

$$h_\ell^{(1)} \sim \frac{1}{z}e^{i(z - \pi\ell/2 - \pi/2)}, \tag{5.84}$$

we obtain the eigenvalue relations

$$-\cot\kappa_1 a \cong \frac{\kappa_0}{\kappa_1} \quad (\ell \text{ even})$$

$$\tan\kappa_1 a \cong \frac{\kappa_0}{\kappa_1} \quad (\ell \text{ odd}). \tag{5.85}$$

In the approximation (5.83), for $n\pi/\xi \cong 1/\sqrt{2}$, *i.e.* n large, the spectrum is approximately given by

$$K_a \cong -\left\{ U - \frac{n_a^2 \pi^2 \hbar^2}{2ma^2} \right\} \tag{5.86}$$

so that

$$E_a \cong Mc^2 - \left(U - \frac{n_a^2 \pi^2 \hbar^2}{2ma^2} \right)$$

$$-\frac{1}{2Mc^2}\left(U - \frac{n_a^2 \pi^2 \hbar^2}{2ma^2} \right)^2 + \cdots. \tag{5.87}$$

The second order relativistic correction to the relativistic spectrum is therefore

$$\frac{\Delta(E_a - Mc^2)}{(E_a - Mc^2)} \cong \frac{1}{2Mc^2}\left(U - \frac{n_a^2 \pi^2 \hbar^2}{2ma^2} \right)^2. \tag{5.88}$$

In (5.86) we see a simple example of the phenomenon described at the end of the previous section. For sufficiently large well-depth U, the eigenvalue K_a can become very large and negative for $U \geq Mc^2$. Thus one cannot argue that the value of K is determined by the aymptotic states; for sufficiently large t there would always be some support for the wavefunction close (Euclidean, *i.e.* with measure d^4x) enough to the light cone to be sensitive to the very large potential, so the idea of a bound state as a composite system of two particles defined asymototically as free would become untenable. For $c \to \infty$ there is no U large enough for this effect to occur, so that the phenomenon is intrinsically relativistic.

For the Coulomb case, the assignment of $K \cong -Mc^2/2$ becomes untenable at

$$Z \geq \frac{M}{\sqrt{M_1 M_2}}(1/\alpha). \tag{5.89}$$

For $M_1 \ll M_2$, for example for one electron in the field of a nucleus, *e.g.* $M_2 = 2ZM_P$, the bound on Z is very high, about 5×10^5. For equal mass, for example two ions, the bound is at $Z \geq 2/\alpha$, which is the order of magnitude of the the value for which the Dirac solutions become unstable.

5.3 The Induced Representation

We have remarked that the solutions of the invariant two body problem results in solutions that are irreducible representations of $O(2,1)$, and not of $O(3,1)$. The ladder representations generated by the action of the Lorentz group on these states cannot be used to obtain representations of the full Lorentz group $O(3,1)$ or its covering $SL(2,C)$ since this construction will connect to the ground state without regularization. Since the differential equations defining the physical states have an operator that is invariant under the action of $O(3,1)$, the function $\Lambda\psi$, for $\Lambda \in O(3,1)$, and ψ a solution, must also be a solution belonging to a representation of $O(3,1)$.

To solve this very fundamental problem, one observes [Arshansky (1989)] that the $O(2,1)$ solutions are constructed in the RMS which is referred to the spacelike z axis. Under a Lorentz boost, the

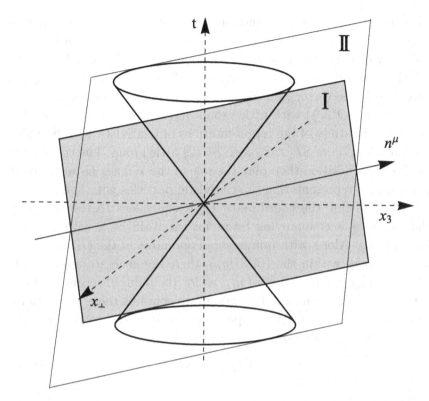

Fig. 3. Orientation of RMS under Lorentz transformation. (Courtesy Springer, Dordrecht).

entire RMS turns, leaving the light cone invariant, as shown in Fig. 3. After this transformation the new RMS is constructed on the basis of a new spacelike direction which we call here m^μ (to distinguish from the timelike n^μ of the induced representation for spin treated in Chapter 4. However, the differential equations remain identically the same in the covariant notation, since the operator form of these equations is invariant. The change of coordinates to RMS variables, although now with new geometrical meaning, has the same form as well, and therefore the set of solutions of these equations have the same structure. These functions are now related to the new "z" axis, and their transformation properties follow the same general rules that we have explained for the induced representations of spin in Chapter 4. Under the action of the full

Lorentz group the wave functions then undergo a transformation involving a linear combination of the set of eigenfunctions found in the previous section; this action does not change the value of the $SU(1,1)$ (or $O(2,1)$) Casimir operator. Together with the change in direction of the vector m^μ, they provide an induced representation of $SL(2,C)$ (or $O(3,1)$ with little group $SU(1,1)$ in the same way that our previous study of the representations of relativistic spin provided a representation of $SL(2,C)$ with $SU(2)$ little group. The coefficients in this superposition then play the role of the Wigner D-functions in the induced representation of relativistic particles with spin.

Let us define the coordinates $\{y_\mu\}$, equivalent to the set $\{x_\mu\}$, defined in an accompanying frame for the RMS(m_μ), with y_3 along the axis m_μ. Along with infinitesimal operators of the $O(2,1)$ generating changes within the RMS(m_μ), there are generators on $O(3,1)$ which change the direction of m_μ; as for the induced representations for systems with spin, the Lorentz group contains these two actions, and therefore both Casimir operators are essential to defining the representations, *i.e.*, both

$$c_1 \equiv \mathbf{L}(m)^2 - \mathbf{A}(m)^2 \tag{5.90}$$

and

$$c_2 \equiv \mathbf{L}(m) \cdot \mathbf{A}(m), \tag{5.91}$$

which is not identically zero, and commutes with c_1.

In the following, we construct functions on the *orbit* of the $SU(1,1)$ little group representing the full Lorentz group; along with the designation of the point on the orbit, labelled by m_μ. These functions constitute a description of the physical state of the system.

It is a quite general result that the induced representation of a noncompact group contains all of the irreducible representations [Mackey (1968)]. We decompose the functions along the orbit into basis sets corresponding to eigenfunctions for the $O(3)$ subgroup Casimir operator $\mathbf{L}(m)^2 \rightarrow L(L+1)$ and $L_1 \rightarrow q$ that take on values that persist along the orbit; these solutions correspond to the *principal series* of Gel'fand [Gel'fand (1963)]. These quantum numbers for the induced representation do not correspond directly to the observed angular momenta of the system. The values that correspond to spectra and wavefunctions with nonrelativistic limit coinciding with those of the nonrelativistic problem, are those with L *half-integer* for the lowest Gel'fand L level. The partial wave expansions in scattering theory, which we discuss in Chapter 6 (for the continuous spectrum of K_{rel}), depend on the quantum number ℓ of the $O(3,1)$ defined on the whole space. These are defined by the quantum form of (5.13), and a magnetic quantum number, which we shall call n, associated with the Casimir of the $SU(1,1)$ discussed above, then playing the role of the magnetic quantum number, as discussed in the previous section for the bound state problem. In fact, in the Gel'fand classification, the two Casimir operators take on the values $c_1 = L_0^2 + L_1^2 - 1$, $c_2 = -iL_0L_1$, where L_1 is pure imaginary and, in general, L_0 is integer or half-integer. In the nonrelativistic limit, the action of the group on the relative coordinates becomes deformed in such a way that the $O(3,1)$ goes into the nonrelativistic $O(3)$, and the $O(2,1)$ into the $O(2)$ subgroup in the initial configuration of the RMS based on the z axis.

The representations that we shall obtain, in the principal series of Gel'fand [Gel'fand (1963)], are unitary in a Hilbert space with scalar product product that is defined by an integration invariant under the full $SL(2,C)$, including an integration over the measure space of $SU(1,1)$, carried out in the scalar product in $L^2(R^4 \subseteq \mathrm{RMS}(m_\mu))$, for each m_μ (corresponding to the orientation of the new z axis, and an integration over the measure of the coset space $SL(2,C)/SU(1,1)$; the complete measure is $d^4yd^4m\delta(m^2 - 1)$, *i.e.*, a probability measure on R^7, where $y_\mu \in \mathrm{RMS}(m_\mu)$. The coordinate description of the quantum state therefore corresponds to an ensemble of (relatively defined) events lying in a set of $\mathrm{RMS}(m_\mu)$s over all possible spacelike $\{m_\mu\}$.

A coordinate system oriented with its z axis along the direction m_μ, as referred to above, can be constructed by means of a coordinate transformation of Lorentz type (here m represents the spacelike orientation of the transformed RMS, not to be confused with a magnetic quantum number),

$$y_\mu = L(m)_\mu{}^\nu x_\nu. \tag{5.92}$$

For example, if we take a vector x_μ parallel to m_μ, with $x_\mu = \lambda m_\mu$, then the corresponding y_μ is λm_μ^0, with m_μ^0 in the direction of the initial orientation of the orbit, say, the z axis. This definition may be replaced by another by right multiplication of an element of the stability group of m_μ and left multiplication by an element of the stability group of m_μ^0, constituting an isomorphism in the RMS.

The variables y_μ may be parametrized by the same trigonometric and hyperbolic functions as in (5.8) since they span the RMS, and provide a complete characterization of the configuration space in the RMS(m_μ) that is universal in the sense that it is the same in every Lorentz frame. It is convenient to define the functions

$$\psi_m(y) = \phi_m(L^T(m)y) = \phi_m(x). \tag{5.93}$$

In a similar way to our previous treatment of the induced representation for spin, we can then define the map of the Hilbert spaces associated with each m_μ in the foliation $\mathcal{H}_m \to \mathcal{H}_{\Lambda m}$ such that the state vectors are related by the norm preserving transformation

$$\Psi_{\Lambda m}^\Lambda = U(\Lambda)\Psi_m. \tag{5.94}$$

In the new Lorentz frame (with $y = L(\Lambda m)x$),

$$
\begin{aligned}
\phi_{\Lambda m}^\Lambda(x) &= {}_{\Lambda m}\langle x|\Psi_{\Lambda m}^\Lambda\rangle \\
&= {}_{\Lambda m}\langle x|U(\Lambda)\Psi_m\rangle = \phi_{\Lambda m}^\Lambda(L^T(\Lambda m)y) \\
&= \psi_{\Lambda m}^\Lambda(y).
\end{aligned}
\tag{5.95}
$$

If $\phi_m(x)$ is scalar under Lorentz transformation, so that (we assume no additional phase)

$$\phi_{\Lambda m}^\Lambda(\Lambda x) = \phi_m(x), \tag{5.96}$$

it follows from (5.95) that

$$U(\Lambda)|x\rangle_m = |\Lambda x\rangle_{\Lambda m}. \tag{5.97}$$

The wave function $\phi_{\Lambda m}^\Lambda(x)$ describes a system in a Lorentz frame in uniform motion with respect to the frame in which the state is described by $\phi_m(x)$, and for which the support is in the RMS($(\Lambda m)_\mu$). The value of this function at x in the new frame is determined by its value at $\Lambda^{-1}x$ in the original frame; moreover, the subensemble associated with values of m_μ over the orbit in the new frame is determined by the subensemble associated with the values of $(\Lambda^{-1}m)_\mu$ in the old frame. We define the description of the state of the system in the new frame in terms of the set (over $\{m_\mu\}$) of transformed wave functions

$$\psi_m^\Lambda(y) \equiv \phi_{\Lambda^{-1}m}(\Lambda^{-1}x)$$
$$= \psi_m^\Lambda(D^{-1}(\Lambda, m)y) \qquad (5.98)$$

where we have used (5.92) (the transformed function has support oriented with m_μ) and defined the (pseudo) orthogonal matrix (we define a "matrix" A as $\{A_\mu{}^\nu\}$)

$$D(\Lambda, m) = L(m)\Lambda L^T(\Lambda^{-1}m). \qquad (5.99)$$

This matrix is analogous to the "little group" acting on the $SU(2)$ of the rotation subgroup discussed in Chapter 4. The transformation $D^{-1}(\Lambda, m)$ stabilizes m_μ^0, and is therefore in the $O(2,1)$ subgroup that leaves the RMS of the original system invariant. (5.98) defines an induced representation of $SL(2, C)$, the double covering of $O(3,1)$.

Classification of the orbits of the induced representation are determined by the Casimir operators of $SL(2, C)$, defined as differential operators on the functions $\psi_m(y)$ of (5.93), *i.e.*, the operators defined in (5.90) and (5.91). To define these variables as differential operators on the space $\{y\}$, we study the infinitesimal Lorentz transformations

$$\Lambda \cong 1 + \lambda, \qquad (5.100)$$

for which

$$\psi^{1+\lambda}{}_m(y) = \psi_{m-\lambda m}(D^{-1}(1 + \lambda, n)y), \qquad (5.101)$$

and λ is an infinitesimal Lorentz transformation (antisymmetric). To first order, the little group transformation is

$$D^{-1}(1 + \lambda, n) \cong 1 - (d_m(\lambda)L(m))L^T(m) - L(m)\lambda L^T(m), \quad (5.102)$$

where d_m is a derivative with respect to m_μ holding y_μ fixed,

$$d_m(\lambda) = \lambda_\mu{}^\nu m_\nu \frac{\partial}{\partial n_\mu}. \qquad (5.103)$$

From the property $L(m)L^T(m) = 1$, it follows that

$$(d_m(\lambda)L(m))L^T(m) = -L(m)(d_m(\lambda)L^T(m)), \qquad (5.104)$$

so that (5.102) can be written as

$$D^{-1}(1 + \lambda, n) \cong 1 + L(m)(d_n(\lambda)L^T(m) - \lambda L^T(m))$$

$$\equiv 1 - G_m(\lambda). \qquad (5.105)$$

For the transformation of ψ_m we then obtain

$$\psi^{1+\lambda}{}_m(y) \cong \psi_m(y) - d_m(\lambda + g_m(\lambda))\psi_m(y), \qquad (5.106)$$

where

$$g_m(\lambda) = G_m(\lambda)_\mu{}^\nu y_\nu \frac{\partial}{\partial y_\mu}. \qquad (5.107)$$

Eq. (5.106) displays explicitly the effect of the transformation along the orbit and the transformation within the little group.

The algebra of these generators of the Lorentz group are investigated in [Arshansky (1989)]; the closure of this algebra follows from the remarkable property of compensation for the derivatives of the little group generators along the orbit (behaving as a covariant derivative in differential geometry). The general structure we have exhibited here is a type of fiber bundle, sometimes called a *Hilbert bundle*, consisting for a set of Hilbert spaces on the base space of the orbit; in this case, the fibers, corresponding to these Hilbert spaces, transform under the little group $O(2,1)$.

There are a set of functions, generated in this way, on the orbit with definite values of the two Casimir operators, as well as $\mathbf{L}(m)^2$ and $L_1(m)$; one finds the Gel'fand Naimark canonical representation with decomposition over the $SU(2)$ subgroup of $SL(2,C)$, enabling an identification of the angular momentum content of the representations [Arshansky (1989)]. With a consistency relation between the Casimir operators (for the solution of the finite set of equations involving functions on the hyperbolic parameters of the spacelike four vector m_μ), we find that we are dealing with the principal series of Gel'fand and Naimark.

In the next chapter we shall discuss scattering theory, and, in particular, the two body case.

Chapter 6

Relativistic Scattering Theory and Resonances

6.1 Introduction

When particles, initially at large distances from each other, enter into a small neighborhood of each other, they exert forces on each other, and eventually move on again to remote regions where they are essentially non-interacting. This configuration is intensively studied, both experimentally and theoretically, in order to obtain information about the structure of the particles and the forces between them.

When the energies of the particles become high, the distance between the particles at closest approach become smaller than at lower energies, and the analysis of the results generally provides more information about the structure of the particles and the forces between them on a smaller scale.

Nonrelativistic scattering theory, which should be valid for particles moving with low energy, is based on the unitary evolution generated by two Hamiltonians; full interacting Hamiltonian of the system, and an unperturbed, or "free" Hamiltonian describing the asymptotic motion when the particles are separated sufficiently to be essentially noninteracting. In this framework, rigorous conditions can be set which can assure the existence of operators, called "wave operators", relating free waves to the physical scattering states and prove asymptotic completeness (for which the range of the wave operators covers the whole space of states); see, for example [Reed (1979), Amrein (1977), Taylor (1972), Newton (1967)]; [Sigal (1987) for the many body case].

When the particles have high energy, relativistic scattering theory must be developed.

Relativistic scattering theory has been generally based on quantum field theory, providing methods of computing an S matrix (transition amplitude operator) by the semi-axiomatic approach of Lehmann, Symanzik and Zimmerman [Lehmann (1955)] or through the use of interaction picture expansion of the perturbed field equations [Schweber(1964), Jauch (1955)], Schwinger-Tomonaga [Tomanaga (1948)]. Feynman's approach to scattering in spacetime [Feynman (1949)], using the method of propagators, is very close to the methods afforded by the covariant quantum theory which we have dicussed in previous chapters, but the notion of invariant evolution is not used explicitly in those computations [Feynman (1950)]. In this chapter, we shall discuss the development of relativistic scattering theory from the first principles of the SHP theory.

A problem common of both the relativistic and nonrelativistic theories has been that of the description of *resonances*, states of matter that are not stable, and for which the apparently irreversible process of decay is of semigroup type (to be defined precisely below). Most of the known particles listed, for example, in the Particle Data Group publications (2014), are so short lived that their time evolution cannot be easily measured directly (their lifetimes are usually estimated by applying the energy time uncertainty relation to the measured widths of the mass distributions), and are considered to be resonances, a scattering in which there is a relatively long time delay [Wigner (1955), Goldberger (1964)]. The time evolution of systems subject to weak decay, such as the free neutron, nuclear beta decay, the muon and K meson systems have been observed, and to very high precision appear to have exponential decay laws. Gamow (1928) proposed the use of a complex energy in the Schrödinger equation to account for the exponential decays observed in nuclear physics, and Wigner and Weisskopf [Weisskopf (1930)] provided a more complete quantum mechanical model which still forms the basis for computations; this theory results in exponential decay for large enough times for single channel decay, but has quadratic time dependence for very short times and does not give semigroup behavior even for intermediate times for decays with several channels (such as the K^0 system [Lee (1956)]).

Lax and Phillips in 1967 [Lax (1967)] presented a description of resonance phenomena with exact semigroup behavior for classical wave equations (such as electromagentic scattering) for which the resonant states are represented as elements of a Hilbert space. The description of Lax and Phillips has recently been formulated in the framework of quantum theory [Strauss (2000a)] thus making the computation of expectation values of observables, as well as the many other properties of resonant quantum states, accessible. As an example, we discuss the application of this method to the relativistic quantum theory [Horwitz (1980)], providing an effective desciption of K^0 meson decay [Strauss (2002)].

6.2 Foundations of Relativistic Scattering Theory

The covariant theory of SHP has a Hamilton-Lagrange formulation of the same form as the nonrelativistic theory, so that the relativistic quantum scattering theory has the same structure as the nonrelativistic theory [Newton (1949), Taylor (1972), Amrein (1977), Merzbacher (1970)]. We shall not, therefore, repeat the procedure and concepts here for the non-relativistic case, but deal directly with the relativistic theory discussed in the previous chapters.

Consider a system characterized by the Hamiltonians K and K_0, corresponding to the full interacting and unperturbed Hamiltonians, *e.g.* for $K = p^2/2M + V$ and $K_0 = p^2/2M$, where $p \equiv p^\mu$ is the momentum four vector of a particle (or the reduced momentum of a two body system) and V is the potential. For the two body problem treated in Chapter 5, K and K_0 may refer to the reduced motion (where M is replaced by the reduced "mass" m used in Eq. (5.6)).

Then there is a state ψ which evolves according the unitary operator $U(\tau) = e^{-iK\tau}$ and an asymptotic state ϕ that evolves according to the unitary operator $U_0(\tau) = e^{-iK_0\tau}$.

The basic condition for scattering theory is that [Jauch (1955)] for every $\epsilon > 0$ there is a T and a ϕ such that

$$\|U(\tau)\psi - U_0(\tau)\phi\| < \epsilon \qquad (6.1)$$

for $|\tau| > T$. Note that this definition does not require that the two terms in (6.1) are identically equal, but only approach each other asyptotically. This formulation permits the establishement of many rigorous properties of scattering systems. Since the norm is invariant under multiplication by a unitary operator, let us multiply both terms by $U^\dagger(\tau) \equiv U^{-1}(\tau)$ to obtain

$$\|\psi - U^\dagger(\tau)U_0(\tau)\phi\| < \epsilon. \tag{6.2}$$

The sequence (6.2) (in the index τ) has the form, for $\{\varphi_n\}$ elements of a dense set,

$$\|\varphi - \varphi_n\| < \epsilon, \tag{6.3}$$

for which $\varphi_n \to \varphi$, so that, if the limit exists,

$$\lim_{\tau \to \pm\infty} U^\dagger(\tau)U_0(\tau)\phi = \psi. \tag{6.4}$$

If the limit exists on a *dense set*$\{\phi\}$ (there are sufficient ϕ's to take limits of the type (6.3) to *any* vector on the Hilbert space), then the operators

$$\Omega_\mp = \lim_{\tau \to \pm\infty} U^\dagger(\tau)U_0(\tau) \tag{6.5}$$

are well-defined. We see this from the fact that the sequence $\|\Omega\phi - \Omega\phi_n\| \leq \|\Omega\|\|\phi - \phi_n\|$. Since Ω is a bounded operator, and ϕ_n may converge to any vector ϕ, then Ω is defined everywhere.

These operators, called the *wave operators* have the property that the full interacting wave function of the system can be expressed in terms of the non-interacting wave function by multiplication by a wave operator, as in (6.4).

Furthermore, by differentiating the operator appearing in (6.5) with respect to τ, one finds, as in the nonrelativistic case, that

$$\lim_{\tau \to \pm\infty} (KU^\dagger(\tau)U_0(\tau) - U^\dagger(\tau)U_0(t)K_0) = 0$$

so that

$$K\Omega_\mp = \Omega_\mp K_0; \tag{6.6}$$

a remarkable property called *intertwining*.

A necessary condition for the existence of the wave operators is that

$$\frac{d}{d\tau}U^\dagger(\tau)U_0(\tau)\phi \to 0,$$

or

$$e^{-iK\tau}(K - K_0)e^{-iK_0\tau}\phi \to 0.$$

Calling $K - K_0 = V$, we see that it is required that

$$\|Ve^{-iK_0\tau}\phi\| \to 0. \tag{6.7}$$

At this point, we notice an essential difference between the relativistic and nonrelativistic theories. If V is a local potential of the form $V(\rho)$ as in the potential models considered in Chapter 5, then we see that the condition (6.7) can be satisfied if the free evolution carries wave packets $\phi(x)$ deep into the spacelike region out of the range of the potential. In the nonrelativistic case, it is easy to see that the free motion can bring the wave packet out of the range of a local potential $V(\mathbf{x})$, but large distances along the hyperbolas $\rho = const$, going asymptotically to the light cone, does not decrease the size of the potential, an effect we have commented on in our previous discussions of the energy spectrum. However, it has been shown that for spacelike momenta, careful estimates, essentially due to a diminishing Euclidean measure of the wave packet as it approaches the light cone where the potential remains large, does admit such a convergence [Horwitz (1980)]. This is an important property for the formulation of relativistic statistical mechanics, a point which we will return to in Chapter 8.

6.3 The S Matrix

The limit (6.2) for $\tau \to -\infty$ corresponds to the transformation of the *in* state to the physical state ψ, and for $\tau \to +\infty$, the transformation from the asymptotic *out* state to ψ also, so that we can write

$$\psi = \Omega_-\phi_{out} = \Omega_+\phi_{in},$$

or

$$\phi_{out} = \Omega_-{}^{-1}\Omega_+\phi_{in}, \tag{6.8}$$

defining the S-matrix (as it is commonly called)

$$S = \Omega_-^{-1}\Omega_+. \tag{6.9}$$

Starting with the Stueckelberg–Schrödinger equation, for $K = K_0 + V$, consider the evolution of the wave function

$$\psi_\tau = e^{-iK_0\tau}\chi_\tau, \tag{6.10}$$

where the functions χ_τ are said to be in the *interaction picture* [Dirac (1947)]. Then,

$$i\frac{\partial}{\partial\tau}\psi_\tau = K_0\psi_\tau + e^{-iK_0\tau}i\frac{\partial}{\partial\tau}\chi_\tau$$

$$= (K_0 + V)\psi_\tau. \tag{6.11}$$

Cancelling $K_0\psi_\tau$ from both sides, we obtain, as in the nonrelativistic theory,

$$i\frac{\partial}{\partial\tau}\chi_\tau = V(\tau)\chi_\tau, \tag{6.12}$$

where $V(\tau) \equiv e^{iK_0\tau}Ve^{-iK_0\tau}$. We can integrate (6.12) from zero to some τ

$$\chi_\tau = \chi_0 - i\int_0^\tau V(\tau')\chi_{\tau'}d\tau' \tag{6.13}$$

and, in case V is small [Kato (1980)], iterate to get a convergent series expansion for an evolution $U(0,\tau)$ for the interaction picture states χ_τ. In the first iteration, one replaces χ_τ' in the integrand by the form given by (6.13), with integration (on, say, τ'' running up to τ'). Thus succesive iterations contain integrations up to the previous time τ, and the result is a sum of τ-ordered integrals. As in the nonrelativistic theory [Amrein (1977)], the integrals can be formally completed to the endpoint τ, after dividing by $n!$ in each nth iterate, with the well-known (in nonrelativistic scattering theory) result

$$U(0,\tau) = \left(e^{-i\int_0^\tau V(\tau')d\tau'}\right)_+, \tag{6.14}$$

where the $+$ subscript implies τ-ordering in the series expansion. The starting point $\tau = 0$ is arbitrary, and the definition can be extended to $U(\tau_1, \tau_2)$. These operators satisfy

$$U(\tau_1, \tau_2)U(\tau_2, \tau_3) = U(\tau_1, \tau_3). \qquad (6.15)$$

Comparison of the definition of χ_τ with the definition of the wave operator (7.10)[1] shows that in fact

$$\lim_{t \to +\infty} U(0, \tau) = \Omega_-, \qquad (6.16)$$

and the S matrix is then given by

$$S = \lim_{\tau_1 \to \infty, \tau_2 \to -\infty} U(\tau_1, \tau_2). \qquad (6.17)$$

There is an alternative form for understanding the wave operators and their physical properties associated with the Green's function [Taylor (1972), Amrein (1977), Newton (1967)] (often called "resolvent" in the mathematical literature).

Let us define for the relativistic theory the unperturbed and perturbed Green's functions

$$\begin{aligned} G^0(z) &= (z - K_0)^{-1} \\ G(z) &= (z - K)^{-1}, \end{aligned} \qquad (6.18)$$

where z may be real or complex (in the upper half plane, as we shall see below). In the nonrelativistic case, where the spectrum of H (and H_0) is often bounded from below, the operators K and K_0 are generally not bounded from below (due to the hyperbolic differential operator for the free motion $p^\mu p_\mu$). However, in the reduced two body problem with symmetric potential, as we have seen in Chapter 5, the reduced Hamiltonian K_{rel} is bounded from below, providing the Green's function for the relative motion $G(z)$ (as well as the scattering operator $T(z)$ to be defined in (6.19)) with simple properties for analytic continuation and causal structure, relevant to the properties of resonances, to be discussed in a later section. Although we review the construction in the following, the

---

[1]Since
$$\chi_\tau = e^{iK_0\tau}e^{-iK\tau}\psi = U(\tau, 0)\psi,$$
as $\tau \to \infty$, this becomes $\Omega_-^\dagger \psi.$, *i.e.*, $U(\tau, 0) \to \Omega_-^\dagger.$

development of the formal scattering theory is almost exactly the same as in the nonrelativistic theory [Amrein (1977), Taylor (1972), Newton (1967)] primarily due to the fact that our formulation of relativistic quantum theory admits a Hamiltonian type structure.

For the scattering problem it is convenient to define another operator, called the "T-matrix", by

$$T(z) = V + VG(z)V, \qquad (6.19)$$

which has the same analytic properties as $G(z)$. Multiplying (6.19) by G_0, we obtain

$$G_0 T(z) = G_0 V + G_0 V G(z)V,$$

but since, by (6.18), it is an identity (sometimes called the second resolvent equation) that

$$G = G_0 + G_0 V G, \qquad (6.20)$$

it follows that

$$G_0(z)T(z) = G(z)V. \qquad (6.21)$$

Multiplying on the right by G_0 one finds, similarly,

$$T(z)G_0(z) = VG(z). \qquad (6.22)$$

A useful integral equation for the T operator can be obtained from (6.20) and (6.22); replacing VG in (6.20) by TG_0 as in (6.22), we obtain

$$G(z) = G_0(z) + G_0(z)T(z)G_0(z). \qquad (6.23)$$

Therefore, the information contained in $T(z)$ (on the effect of interaction) is equivalent to that of $G(z)$.

Furthermore, if we replace GV in the definition of the T operator (6.19) by $G_0 T$ as in (6.21), we obtain

$$T(z) = V + VG_0(z)T(z). \qquad (6.24)$$

This equation is known as the Lippmann-Schwinger equation (1950) for the computation of $T(z)$, and used in many applications (*e.g.* Adler-Weisberger sum rule (1950)). If, for example, V is very small, one obtains the lowest Born approximation (we shall see below that T is directly related to the scattering transition amplitude in the form $S = 1 - 2\pi i T$) for which $T \sim V$, and in general, by iteration, one obtains the full Born series:

$$T(z) = V + VG_0V + VG_0VG_0V + \cdots \qquad (6.25)$$

It is interesting to note that, as is well known for the nonrelativistic theory, $G(z^*) = G(z)^\dagger$, and therefore $T(z^*) = T(z)^\dagger$, *i.e.*, the Hermitian conjugate, well defined in the SHP Hilbert space of states.

We now establish important connections between the Green's function, the T-matrix, and the S matrix. The limit (6.5) for the definition of the wave operator may be written in a different and useful way as (we write the result first for Ω_-)

$$\Omega_- = \lim_{\tau \to \infty} U^\dagger(\tau)U_0(\tau) = \lim_{\epsilon \to 0} \int_0^\infty d\tau e^{-\epsilon\tau} U(\tau)^\dagger U_0(\tau), \qquad (6.26)$$

with the limit in ϵ taken from above. This result, a procedure introduced by Abel [Amrein (1977)], is easily proved by assuming that there is a number T such that for $\tau > T$ the product of unitaries under the limit has converged. The finite part of the integral vanishes when $\epsilon \to 0$. What remains is then an integral of the exponential from T to ∞; this cancels the factor ϵ and leaves unity in the limit ϵ goes to zero. To work with this formula, we note that the prefactor of ϵ can be provided by differentiating the exponent with a minus sign, and then integrating by parts, one finds

$$\Omega_- = 1 + i \lim_{\epsilon \to 0} \int_0^\infty d\tau e^{-\epsilon\tau} U(\tau)^\dagger V U_0(\tau). \qquad (6.27)$$

In a similar way, one obtains

$$\Omega_+ = 1 - i \lim_{\epsilon \to 0} \int_{-\infty}^0 d\tau e^{+\epsilon\tau} U(\tau)^\dagger V U_0(\tau). \qquad (6.28)$$

Now, let $\Omega_\mp$ act on ϕ respectively, for ϕ_{in} and ϕ_{out}, represented as

$$|\phi\rangle = \int d^4p |p\rangle \langle p|\phi\rangle \qquad (6.29)$$

In its action on this state, in this representation, the operator K_0, which may depend on p^μ alone, takes on the value K_p. There are cases for which K_0 may be more involved; in such cases, one uses its spectral representation (usually absolutely continuous). Carrying out the integration over τ, and using the definition (6.18) for the Green's function, we then have

$$\Omega_-|\phi\rangle = |\phi\rangle + \lim_{\epsilon\to 0}\int d^4p\, G(K_p - i\epsilon)V|p\rangle\langle p|\phi\rangle \qquad (6.30)$$

and

$$\Omega_+|\phi\rangle = |\phi\rangle + \lim_{\epsilon\to 0}\int d^4p\, G(K_p + i\epsilon)V|p\rangle\langle p|\phi\rangle. \qquad (6.31)$$

Note that the labels on $\Omega_\pm$ go with $\pm i\epsilon$ in these expressions. We also remark that the incoming or outgoing wave packets associated with these states may be moving forward or backward in t as a function of τ (thus the interaction may be capable of inducing pair annihilation, as in Stueckelberg's original conception (see Chapter 1)).

An important result of this computation is that the incoming and outgoing waves are now expressed in terms of the operators GV, enabling us to express, with (6.21), these waves in terms of G_0 and the scattering operator T, *i.e.*,

$$\Omega_-|\phi\rangle = |\phi\rangle + \lim_{\epsilon\to 0}\int d^4p\, G_0(K_p - i\epsilon)T(K_p - i\epsilon)|p\rangle\langle p|\phi\rangle \qquad (6.32)$$

and

$$\Omega_+|\phi\rangle = |\phi\rangle + \lim_{\epsilon\to 0}\int d^4p\, G_0(K_p + i\epsilon)T(K_p + i\epsilon)|p\rangle\langle p|\phi\rangle. \qquad (6.33)$$

Using the same techniques, we now proceed to express the S matrix in terms of the operator $T(z)$ as well. To do this, we write the matrix elements

$$\langle p'|S|p\rangle = \lim_{\tau'\to\infty,\tau\to-\infty}\langle p'|e^{iK_0\tau'}e^{-iK\tau'}e^{iK\tau}e^{-iK_0\tau}|p\rangle;$$

the two limits may be taken simultaneously since both factors are
supposed convergent, so the calculation may be made for

$$\langle p'|S|p\rangle = \lim_{\tau\to\infty} \langle p'|e^{iK_0\tau}e^{-i2K\tau}e^{iK_0\tau}|p\rangle. \tag{6.34}$$

We now, as before, write this as

$$\langle p'|S|p\rangle = \lim_{\epsilon\to 0}\epsilon \int_0^\infty d\tau\, e^{-\epsilon\tau}\langle p'|e^{iK_0\tau}e^{-i2K\tau}e^{iK_0\tau}|p\rangle,$$

and provide the factor ϵ by differentiating the exponent. Integrating
by parts (taking one factor of $e^{-iK\tau}$ to the left and one to the right),
one obtains

$$\langle p'|S|p\rangle = \delta^4(p'-p) - i\lim_{\epsilon\to 0}\int_0^\infty d\tau\Big\{Ve^{i(K_{p'}+K_p-2K+i\epsilon)\tau}$$
$$+ e^{i(K_{p'}+K_p-2iK+i\epsilon)\tau}V\Big\}.$$

Carrying out the integral over τ, one obtains

$$\langle p'|S|p\rangle = \delta^4(p'-p) + \frac{1}{2}\lim_{\epsilon\to 0}\Big\{\langle p'|VG\left(\frac{K_{p'}+K_p}{2}+i\epsilon\right)$$
$$+ G\left(\frac{K_{p'}+K_p}{2}+i\epsilon\right)V|p\rangle. \tag{6.35}$$

Recognizing again that $VG = TG_0$, and replacing K_0 by K_p in the
first term, and $K_{p'}$ in the second, we obtain

$$\langle p'|S|p\rangle = \delta^4(p'-p)$$
$$+ \lim_{\epsilon\to 0}\Big\{\frac{1}{K_{p'}-K_p+i\epsilon} + \frac{1}{K_p-K_{p'}+i\epsilon}$$
$$\langle p'|T\left(\frac{K_{p'}+K_p}{2}+i\epsilon\right)|p\rangle. \tag{6.36}$$

With the property of distributions that

$$\lim_{\epsilon\to 0_+}\left(\frac{1}{x+i\epsilon}-\frac{1}{x-i\epsilon}\right) = -2\pi i\delta(x),$$

we obtain one of the main results of this section, that

$$\langle p'|S|p\rangle = \delta^4(p' - p) - 2\pi i\delta(K_{p'} - K_p)\lim_{\epsilon\to 0}\langle p'|T(K_p + i\epsilon)|p\rangle \quad (6.37)$$

This result is completely analogous to the result of the nonrelativistic formal scattering theory, where the matrix element corresponding to the scattering amplitude is often defined as

$$t(p' \leftarrow p) = \lim_{\epsilon\to 0}\langle p'|T(K_p + i\epsilon)|p\rangle. \quad (6.38)$$

As in the nonrelativistic case, it follows from (6.24) that, although the physically relevant value of $T(z)$ is the so-called "on-shell" value of the limit for $z \to K_p + i0$, the matrix elements $\langle p'|T(z)|p\rangle$ satisfy the integral Lippmann-Schwinger [Lippman (1950)] equation

$$\langle p'|T(z)|p\rangle = \langle p'|V|p\rangle + \int d^4p'' \frac{\langle p'|V|p\rangle}{z - K_{p''}}\langle p''|T(z)|p\rangle. \quad (6.39)$$

We shall discuss in the next section how the cross section is computed from this amplitude, yielding somewhat different content from the nonrelativistic case, and then study the representation of the scattering amplitude in terms of the analog of the Bessel Legendre expansion of the nonrelativistic theory.

6.4   Cross Sections

The computation of the cross sections in the relativistic SHP scattering theory was made by Lavie and Horwitz (1982). The probability for the scattered particle (event) to lie in a region d^4p is given by

$$w(d^4p \leftarrow \phi_{in}) = |\phi_{out}{}^{scatt}|^2 d^4p, \quad (6.40)$$

where, from the relation (6.8), determining the out-wave from the in-wave through the S-matrix, we recognize from (6.37) that the *scattered* part of the out-wave $\phi_{out}{}^{scatt}$ is constructed by means of the T matrix. Assuming that the laboratory detectors are sensitive to the *direction* of the momentum **p** and not its magnitude, we integrate over all $|\mathbf{p}|$ to obtain the probability to find the particle emerging with energy dp^0 around p^0, and three momentum in the solid angle $d\Omega$ around $\mathbf{p}$,

$$w(d\Omega dp^0 \leftarrow \phi_{in}) = d\Omega dp^0 \int d|\mathbf{p}||\mathbf{p}|^2 |\phi_{out}^{scatt}(\mathbf{p}, p^0)|^2. \qquad (6.41)$$

Note that the energy p^0 is held fixed for the integration, and, since $p^0 = \sqrt{\mathbf{p}^2 + m^2}$, if the particle remains close to "mass shell", the integration over the magnitude of the momentum is restricted to a small range.

We have so far taken into account a single incoming wave packet. For a beam of wave packets at impact parameters ρ distributed over times $\{x_0\}$ in a pulse in the beam, the total number of scatterings into $d\Omega dp^0$ would be

$$N_{scatt}(d\Omega dp^0) = \int d^3\rho \int dx^0 w(d\Omega dp^0 \leftarrow \phi_{in}^{\rho,x^0}) n_{inc}(\rho, x^0), \qquad (6.42)$$

where $n_{inc}(\rho, x^0)$ is the number of packets per unit area and unit time perpendicular to the motion of the beam. Since the beam should cover the potential, the integrals can be extended to infinity without changing the result. For n_{inc} constant, the cross section can then be defined by

$$\sigma(d\Omega dp^0 \leftarrow \phi_{in}) = \frac{N_{sc}(d\Omega dp^0)}{n_{inc}} = \int d^3\rho \int dx^0 w(d\Omega dp^0 \leftarrow \phi_{in}^{\rho,x^0}). \qquad (6.43)$$

This definition is given in terms of a *number* divided by a *density* (equivalent to a rate divided by a flux on $\Delta\tau$), and has dimension three, [Cook (1957)] has given a similar definition). In four dimensions, we see that, as in three dimensions, the dimension of the cross section is the dimension of the space minus the one dimension of the beam. The extra dimension in the cross section can be understood in terms of the time interval which spans the extension of the potential in the relative time (which may be, for example, of the order of the spatial range divided by c). This factor of the time interval emerges explicitly in the nonrelativistic limit, where the scattering amplitude contains a delta function $\delta(p^0 - \bar{p}^0)$, where $\bar{p}^0$ is the average energy of the packet, a result of integration over x^0 required to sample the potential through its spread in t [Horwitz (1982)].

As in the nonrelativistic theory, an optical theorem can be proven, relating the total cross section to the imaginary part of the forward cross section, as a result of the unitarity of the S matrix. See [Horwitz (1982)], where the Feynman rules, electromagnetic scattering, and many body scattering are discussed, and the Rutherord cross section is obtained.

6.5 Two Body Partial Wave Analysis

In this section, we use the framework established in Chapter 5 to describe scattering on the elementary level defined by the partial wave expansions of the nonrelativistic theory. The states of the two body systems are given in terms of the RMS structure applicable to the bound states in the part of the spectrum above the ionization point.

The partial wave expansion that we shall discuss contains phase shifts labelled by the quantum number ℓ which determines the value of the $O(3, 1)$ Casimir operator intrinsic to the RMS, and corresponds to the nonrelativistic orbital angular momentum quantum number. As we have seen in the previous section, the hyperarea in four-dimensional spacetime perpendicular to the space direction of the incident beam is three dimensional (L^2T). This cross section would include the scattering of an ensemble of events that includes all possible distributions of β, and therefore to possible mass changes of the particles after the scattering. Restricting to a small neighborhood of permissible β's corresponds to a restriction to a definite mass shift; the result carries a Jacobian $dx^0/d\beta$, reducing the dimensionality of the incident flux to that of current per unit area, thus accounting for the factor of time required to cover the action of the potential discussed in the previous section. It is interesting to note that the time dimension of the cross section formula corresponds in this sense to the inclusion of inelastic phenomena (consistent with the $\Delta t \Delta E$ relation).

It will be recalled that the representations of $O(3,1)$ provided by the solution for the bound state problem constituted an induced representation on an orbit over spacelike directions. In this construction, the wave function carries a representation of $O(2,1)$ that moves along an orbit labelled by the spacelike vector m_μ, accompanied by an $O(2,1)$ Wigner "rotation". The structure of this motion was analyzed [Arshansky (1989)] into irreducible representations of $O(3) \subset O(3,1)$ with quantum numbers (L,q), thus obtaining the principal series of Gel'fand [Gel'fand (1963)]. The bound states can be described equally well for any choice of m_μ; the mass levels are completely degenerate with respect to this choice. In the problem of two-body scattering, however, the direction of the beam selects a definite spacelike direction. For the wavefunction with m_μ component oriented along this direction, one can argue that the scattered wave should be maximally symmetric about this axis. This maximally symmetric state is the one for which the Gel'fand representation contains only the value of L corresponding to the lowest weight of the principal series, and we shall assume that the scattering matrix (which is diagonal in m_μ) is described, to a good approximation, by such a state [Arshansky (1989)]. The result we shall obtain in this way agrees in form with the well-known partial wave expansion in the non-relativistic theory. Alternative choices of m_μ, resulting in states with less symmetry, evidently do not contribute in an important way to the partial wave expansions which have been useful in describing scattering experiments.

We therefore take the general form of the two body wave function corresponding to a definite value c_2 for the second Casimir operator $\mathbf{L} \cdot \mathbf{A}$ and a definite direction for m_μ the form

$$\psi_{n_\mu}^{c_1} = \frac{1}{\sqrt{\rho \sin\theta \cosh\beta}} \Sigma_{\ell,n,k,L,q} A_{\ell n k}^{Lq} \hat{R}_\ell^\kappa(\rho) \hat{\Theta}_\ell^n(\theta)$$

$$\times \Xi_{n,k}^{c_2,L}(u) P_{q-M_k}^L(z) e^{-iq\gamma} \hat{\chi}_{n+l}^{-n}(\beta,\varphi), \qquad (6.44)$$

where $u = \tanh\alpha$, $z = \sin\omega$, and α, ω, γ are the angles and hyperangle representing the orientation of the spacelike vector m_μ, and the functions appearing on the the right hand side of (6.44) are defined in Arshansky (1989). The variables $\rho, \theta, \beta, \varphi$ correspond to the relative coordinates in the RMS defined by m_μ. The measure on the Hilbert space $\mathcal{H}_m$ to which these functions belong is $d\mu = \rho^2 \sin^2\theta \cosh\beta d\rho d\beta d\theta d\phi$. The integer parameter n, determining the Casimir operator for the $O(2,1)$ little group, plays the role of the magnetic quantum number in the corresponding nonrelativistic problem; here, it fixes the relation between the value of the first Casimir operator $\mathbf{L}^2 - \mathbf{A}^2$ and c_2 according to $(\hat{n} = n + \frac{1}{2})$

$$-c_1 = 1 - \hat{n}^2 + c_2^2/\hat{n}^2, \tag{6.45}$$

the consistency relation found in Chapter 5.

The differential equations for the functions $\hat{R}_\ell^\kappa(\rho), \hat{\Theta}_\ell^n(\theta)$ obtained from separation of variables related to the accompanying coordiates y^μ, are identical to the equations satisfied by the corresponding nonrelativistic functions (the "hats" denote the extraction of factors $1/\sqrt{\rho}$ and $1/\sqrt{\sin\theta}$ from the functions obtained from the relativistic equations); the extra factors are included explicitly. We have also extracted the factor $1/\sqrt{\cosh\beta}$ in the function $\hat{\chi}_{n+l}^{-n}(\beta,\varphi)$, constituting the irreducible representations of $O(2,1)$.

These functions are of quite a different form from those of the usual partial wave expansion; if, however, we choose m_μ to be directed parallel to the incoming beam, which we take to be the z axis, the parameters α, ω and γ are zero, and the accompanying coordinates y^μ coincide with the base relative coordinates x^μ for this orientation. Only the lowest weight of the Gel'fand representation, $L = \frac{1}{2}$ and $n = k = 0$ contribute. We are therefore left with the simple form

$$\psi(x) = \Sigma_{\ell=0}^\infty \frac{A_\ell \hat{R}_\ell^\kappa(\rho) P_\ell(\cos\theta) e^{i\varphi/2}}{\sqrt{\rho \sin\theta \cosh\beta}}, \tag{6.46}$$

very similar to the usual partial wave expansion. The half-integer phase factor, discussed above in connection with the bound state functions, is a particular peculiarity associated with the topology of the RMS, but does not influence the experimental predictions of the scattering theory on this level.

The coefficients A_ℓ can be determined, as for the usual partial wave expansion for the nonrelativistic problem, by requiring that ψ

take the form of the asymptotic incoming wave ($\tau \to -\infty$ for a wave packet on the value κ of the z-component of momentum),

$$\psi_{inc} \sim \frac{e^{i\kappa\rho\cos\theta}e^{i\varphi/2}}{\sqrt{\rho\cosh\beta\sin\theta}}, \tag{6.47}$$

where we have used the fact that $\hat{R}_\ell^\kappa(\rho)$ is a solution of the nonrelativistic radial equation in ρ, and that $\rho \to \infty$ spacelike in this limit. The solution (6.47) is an eigenstate of the four momentum operator p_μ with eigenvalue κ for p_3; asymptotically, the other components vanish, to that $p_\mu \sim (0, 0, 0, \kappa)$.

The *conserved* current associated with the wave function (6.46) (it is this *relative* current which is associated with the counting of scattering events) is given by

$$j_\mu = -\frac{i}{2m}(\psi^*(x)\partial_\mu\psi - \partial_\mu\psi^*\psi). \tag{6.48}$$

The τ integration that is required to convert the τ-dependent current density into a conserved current in the asymptotic free particle case, serves to link the mass squared values κ^2 in the two factors ψ with a δ-function. In an interval $d\kappa^2/2\pi$, one obtains (the other components vanish)

$$j_z = \frac{\kappa}{m}\frac{\sin\theta\cosh\beta}{\rho}. \tag{6.49}$$

The relation between the parameters β and θ determines the synchronization between the pair of events being considered, and this determines the mass change during the scattering.

The function $\hat{R}_\ell^\kappa(\rho)$, as remarked above, is a solution of the nonrelativistic radial equation, and therefore results in an outgoing wave of the form

$$\psi^{(+)} \sim \frac{e^{i\varphi/2}}{\sqrt{\rho\cosh\beta\sin\theta}}\left\{e^{i\kappa\rho\cos\theta} + \frac{1}{\rho}f(\theta)e^{i\kappa\rho}\right\}, \tag{6.50}$$

where, in the Legendre expansion of $f(\theta)$ the coefficients, following the usual arguments [Merzbacher (1970)] for asymptotic values of ρ, are related to a set of phase shifts $\delta_\ell(\kappa)$ according to

$$f(\theta) = \frac{1}{2i\kappa}\Sigma_{\ell=0}^\infty(2\ell + 1)(S_\ell - 1)P_\ell(\cos\theta), \tag{6.51}$$

and

$$S_\ell = e^{2i\delta_\ell}. \tag{6.52}$$

The quantities S_ℓ are the ℓ components of the S matrix. The set of numbers δ_ℓ are the same as the nonrelativistic phase shifts (as functions of κ, however) since the radial equation (for $V(\rho)$ the same form as the nonrelativistic $V(r)$) is identical to that of the nonrelativistic problem.

Using the second part of (6.48) and (6.50) we may compute the outgoing current (derivatives with respect to θ, β and φ contain factors that go like $1/\rho$ and vanish asymptotically). The contributions of the ρ derivatives multiplied by the four-volume element divided by $d\rho$ (the infinitesimal volume element lying on the constant ρ hypersurface) correspond to the number of particles per unit time scattered through this surface element associated with the currents (6.48) in the outgoing wave. For the part of this flow through a surface element normal to the scattering direction specified by unit vector x^μ/ρ, one obtains $(x^\mu/\rho)j_\mu^{scatt}$ for the number of particles per unit time through this surface element. Dividing by the incident flux (and integrating over the azimuthal angle φ), we obtain the differential cross section

$$d\sigma(\theta) = 2\pi|f(\theta)|^2 d\Omega(\theta), \tag{6.53}$$

where

$$d\Omega(\theta) = \sin\theta d\theta. \tag{6.54}$$

As we have remarked above, the formulas for scattering we have obtained apply both to elastic and inelastic (mass changing) scattering.

6.6 Unitarity and the Levinson Theorem

The total probability for the incoming wave (6.47) is

$$\int \rho^3 \sin^2\theta \cosh\beta d\rho d\theta d\varphi |\psi_{inc}|^2 = \int_0^R d\rho = R, \tag{6.54}$$

for an interval $d\beta$, as for our computation of the current above. In the computation of the norm of the outgoing wave $\psi^{(+)}(x)$ of (6.50) in this interval $d\beta$, with the help of the orthogonality relations for Legendre polynomials, one finds precisely the result (6.54). This demonstrates unitarity of the S matrix for *each* value of β, *i.e.*, for each state of inelasticity described by this scattering system. It furthermore follows from the form of $f(\theta)$ given in (6.51) that the optical theorem follows in the usual form.

The analytic properties of $S_\ell(\kappa)$ follow from the radial equation and the asymptotic form (6.50) for the outgoing wave. Following Levinson (1949) we identify the part of the wave function with asymptotic behavior $\sim \exp(+i(\kappa\rho - \pi/2))$; the limit of this function for $\rho \to 0$ (on the light cone) is called $D_\ell(\kappa)$; then, $D_\ell^*(\kappa) = (-1)^\ell D_\ell(\kappa)$. Since

$$S_\ell(\kappa) = D_\ell^*(\kappa)/D_\ell(\kappa), \qquad (6.55)$$

integration on κ from $-\infty$ to $+\infty$ (noting that $\delta_\ell(\kappa) = -\delta_\ell(-\kappa)$, one finds that

$$\delta_\ell(\infty) - \delta_\ell(0) = -\pi N_b^\ell, \qquad (6.56)$$

where N_b^ℓ is the number of bound states for a given ℓ. This connection between the number of bound states and the scattering phase shifts is consistent with our formulation of scattering in the RMS. Moreover, the bound states with support in a given RMS are associated with the scattering for the same direction of m_μ of that RMS.

We note from this analysis that when δ_ℓ goes through the value $\pi/2$ the cross sention goes through a maximum. This fact is associated with the interpretation of a *resonance* for this value of κ, corresponding to a complex pole in the lower half plane of the S matrix.

6.7 Resonances and Semigroup Evolution

We have introduced the Green's function in (6.18) as an important constituent in the development of formal scattering theory. This function also arises in the Laplace transform of the Stueckelberg–Schrödinger equation (and thus becomes associated with the idea of propagation). As for the nonrelativistic Schrödinger equation, for τ independent generator K, the formal integrated solution is

$$\psi_\tau = e^{-iK\tau}\psi_0 \qquad (6.57)$$

Thus the unitary evolution generated by K "propagates" the wave function forward in τ, but not necessarily in t. Thus the spacetime diagrams of Feynman, as he pointed out in his paper [Feynman (1949)], can be understood as lines propagating forward and backward in time according to this evolution; in interaction picture, the straight lines correspond to unperturbed propagation, and the vertices, to some interaction carried by the full Hamiltonian K. The standard interaction picture applied to this evolution can be represented in terms of such diagrams (as will be pointed out in Chapter 7, for gauge fields, the fifth field induced by gauge invariance of the Stueckelberg–Schrödinger equation must be taken into account).

In the following, we make a connection between the resonances which occur in scattering theory, recognized as fairly well defined maxima in the cross sections defined by the scattering amplitude $t(p' \leftarrow p)$, corresponding to values of δ_ℓ defined in (6.51) and (6.52) going through a value of $\pi/2$ (or multiples of it), and the description of unstable systems [Horwitz (1971)].

In 1926, for the description of the decay of an unstable system in the nonrelativistic theory, George Gamow (1928) suggested that the Schrödinger equation be assigned a complex eigenvalue, $E - i\frac{\Gamma}{2}$ so that the solution would gave an exponential decay law. The idea was not thoretically tenable since the momentum is proportional to the square root of the energy in nonrelativistic mechanics, and this would imply a complex momentum and associated instabilty of the spatial wave function. Wigner and Weisskopf in 1930 [Weisskopf (1930)] proposed a model in which the initial state of a quantum mechanical system is considered the "particle", or the state of the unstable system, and its projection back into that state after time t would be understood as the amplitude for the particle to remain in its initial state, *i.e.* the "survival probability amplitude" would be[2]

$$A(t) = (\psi, e^{-iHt}\psi). \tag{6.58}$$

In the relativistic theory we may introduce the analogous definition,

$$A(\tau) = (\psi, e^{-iK\tau}\psi), \tag{6.59}$$

where, for potential type problems, K may be taken as the reduced motion which, for the two body problem, may be bounded from below.[3]

The Laplace transform of (6.59) provides a useful [Horwitz (1971)] interpretation of the Green's function introduced in (6.18), *i.e.*

$$A(z) = \int_0^\infty e^{iz\tau} A(\tau)$$

$$= i\left(\psi, \frac{1}{z-K}\psi\right), \tag{6.60}$$

is well-defined (and analytic) for $Im z > 0$, exhibiting explicitly the Green's function, now understood in terms of a "propagator", as the Laplace transform of the unitary evolution.

The inverse Laplace transform is an integral to be carried out on a line in the complex z plane just above the real axis (where the function is analytic) from $+\infty$ to $-\infty$,

$$A(\tau) = \frac{1}{2\pi i} \int_{\infty+i\epsilon}^{-\infty+i\epsilon} dz\, e^{-iz\tau} A(z). \tag{6.61}$$

If the spectrum of K (or of H in the nonrelativistic case), for, *e.g.*, the reduced motion of a two body system, runs from zero to infinity, then the integration on the negative real axis can be moved to the

---

[2]It will be convenient in the remainder of this chapter to use the round bracket for scalar products when using normalized functions, and the angular bracket for non-normalized generalized states.

[3]The discussion which follows can be applied directly to the nonrelativistic case by transposing τ to t. We have maintained the notation τ here to be consistent with the relativistic framework.

lower half plane, running along the imaginary axis, where the value is suppressed by the exponential $\exp(-iz\tau)$; the contributions are very small for the interval sufficiently below the branch point, and τ sufficiently large. The integral along the positive real axis can be lowered to the second Riemann sheet by considering the difference, with the help of the identity given after (6.36)

$$(\psi, G(\kappa + i\epsilon)\psi) - (\psi, G(\kappa - i\epsilon)\psi) = 2\pi i \chi(\kappa), \qquad (6.62)$$

where $\chi(\kappa)$ is the spectral weight factor of the expectation value of the spectral representation of the operator $(z - K)^{-1}$. The term $(\psi, G(\kappa - i\epsilon)\psi$ is evaluated by analytic continuation around the real axis to a point just below the positive real axis.[4] Rotating the integral on the positive real axis into the lower half plane (second sheet), one finds that there can be contributions from singularities in the lower half plane. In the case of a pole, which one might understand as the remnant of a bound state, a pole on the real axis, pulled down into the lower half plane by the interaction, the passage of the contour of integration over this pole extracts a residue proportional to $e^{-iz_P\tau}$, which may dominate the entire integral (for times not too long and not too short), in agreement with the proposal of Gamow. For very short times this expresssion will not be dominant, and for very long times, the contribution of the branch cut dominates, for which there may be a polynomial type decay law [Bleistein (1977)].

Furthermore, the pole contribution does not correspond to any physical state. It has been shown, on the other hand, that there is a vector in a Banach space (an element of a Gel'fand triple constructed in the dual to a subspace of the original Hilbert space [Horwitz (1978), Baumgartel (1976)]) that can be constructed to correspond to this pole with exact exponential decay, but it is difficult to interpret such a construction as a physically meaningful state; expectation

---

[4]For the case of a continuous spectrum from $-\infty$ to $+\infty$, two functions, one analytic in the upper half place, and the other in the lower half plane, according to forward and backward evolution may be defined, and a similar method may be applied by moving the integral along the real line in the upper half plane into the lower half plane.

values and scalar products would not be generally defined (see *e.g.* [Bohm (1989)]).

The T matrix, (6.19), contains this function as well, playing an essential role in the construction of the S matrix and the scattering amplitudes for $Imz \to 0_+$, as in (6.38).

Although this formula may provide an exponential behavior for $|A(\tau)|^2$ for sufficiently long (but not too long) times, as we have seen, for very short times it generally displays a decay law which is not consistent with the exponential form. For short times,

$$A(\tau) = 1 - i\langle K \rangle \tau - \frac{\langle K^2 \rangle}{2}\tau^2 + \cdots . \qquad (6.63)$$

It then follows that

$$|A(\tau)|^2 = 1 - \Delta K^2 \tau^2 + \cdots , \qquad (6.64)$$

with $\Delta K^2 = \langle K^2 \rangle - \langle K \rangle^2$, not consistent with the semigroup property.

On the other hand, the *semigroup* law of evolution, expected from reasonable arguments to be valid for irreversible processes, such as the decay of an unstable system,[5] is defined by the relation

$$Z(\tau_1)Z(\tau_2) = Z(\tau_1 + \tau_2), \qquad (6.65)$$

for $\tau_1, \tau_2 \geq 0$; $Z(\tau)$ has no inverse, unlike a one parameter group, such as the unitary Stueckelberg–Schrödinger evolution $e^{-iK\tau}$, which has the property (6.65) for all τ_1, τ_2. The model proposed by Gamow, for which $|A(t)|^2 \propto e^{-\Gamma t}$, does have this property, in agreement with experiment, but the derivative of this function at zero is $\propto -\Gamma$, so the function approaches unity at zero linearly, not quadratically, as it would for almost any Hamiltonian (with finite dispersion in state ψ) in the Wigner–Weisskopf model. One can argue that in many cases the very short time before an approach to exponential [Misra (1977)] behavior would not be observable experimentally,

---

[5]Based on the argument that one can stop the evolution at any moment and then proceed as if starting from the new initial conditions, with a result equivalent to letting the system develop undisturbed for the entire time, an essentially Markovian hypothesis.

and this has justified its use in many cases, but the fact that the evolution is not semigroup has consequences for the application of the idea to two or more dimensions, such as for the neutral K meson decay, where it has been shown to be quantitatively inapplicable [Cohen (2011)]. One finds that the poles of the resolvent for the Wigner–Weisskopf evolution of the two channel system results in non-orthogonal residues that generate interference terms, which make the non-semigroup property evident even for times for which the pole approximation is valid, a domain in which exponential decay for the single channel system is very accurately described by the Wigner–Weisskopf model.

The Yang-Wu [Wu (1975)] parametrization of the K^0 decay processes, is based on a Gamow type evolution generated by an effective 2×2 non-Hermitian matrix Hamiltonian, on the other hand, results in an evolution that is an exact semigroup. This parametrization consistent to a high degree of accuracy with the experimental results on K-meson decay (Data (2014)). It does not, however, have a consistent theoretical structure based on first principles.

We shall discuss below a fundamental theory (based on the work of Lax and Phillips [Lax (1967)]) in which the evolution law is precisely semigroup and identifies the resonance with a quantum state, and discuss how it can be applied to the relativistic evolution of the neutral K meson system, explaining as well the origin of the phenomenological parametrization of Yang and Wu.

6.8 Lax Phillips Theory

The quantum Lax-Phillips theory [Strauss (2000a)] which we discuss in the following, constructed by embedding the quantum theory into the original Lax-Phillips scattering theory [Lax (1967)] (originally developed for hyperbolic systems, such as acoustic or electromagnetic waves), describes the resonance as a *state* in a Hilbert space, and therefore it is possible, in principle, to calculate all measurable properties of the system in this state. Moreover, the quantum Lax-Phillips theory provides a framework for understanding the decay of an unstable system as an *irreversible* process. It appears, in fact, that this framework is categorical for the description of irreversible processes for the evolution of an "isolated" quantum system.

The scattering theory of Lax and Phillips [Lax (1967)], originally developed for the description of resonances in classical wave problems such as electromagentic or acoustic, assumes the existence of a Hilbert space $\overline{\mathcal{H}}$ of physical states in which there are two distinguished orthogonal subspaces $\mathcal{D}_+$ and $\mathcal{D}_-$ with the properties

$$U(\tau)\,\mathcal{D}_+ \subset \mathcal{D}_+ \qquad \tau > 0$$
$$U(\tau)\,\mathcal{D}_- \subset \mathcal{D}_- \qquad \tau < 0$$
$$\bigcap_\tau U(\tau)\,\mathcal{D}_\pm = \{0\}$$
$$\overline{\bigcup_\tau U(\tau)\,\mathcal{D}_\pm} = \overline{\mathcal{H}}, \tag{6.66}$$

i.e., the subspaces $\mathcal{D}_\pm$ are stable under the action of the full unitary dynamical evolution $U(\tau)$, a function of the physical laboratory time, which we identify with the universal invariant time τ of the Stueckelberg theory discussed above for its application to the relativistic theory. Over all τ, the evolution operator generates a dense set in $\overline{\mathcal{H}}$ from either $\mathcal{D}_+$ or $\mathcal{D}_-$. We shall call $\mathcal{D}_+$ the *outgoing subspace* and $\mathcal{D}_-$ the *incoming subspace* with respect to the group $U(\tau)$.

A theorem of Sinai [Cornfield (1982)] assures that $\overline{\mathcal{H}}$ can be represented as a family of Hilbert spaces obtained by foliating $\overline{\mathcal{H}}$ along a real line, which we shall call $\{s\}$, in the form of a direct integral

$$\overline{\mathcal{H}} = \int_\oplus \mathcal{H}_s, \tag{6.67}$$

where the set of auxiliary Hilbert spaces $\mathcal{H}_s$ are all isomorphic. Representing these spaces in terms of square-integrable functions, we define the norm in the direct integral space (we use Lesbesgue measure) as

$$\|f\|^2 = \int_{-\infty}^\infty ds \|f_s\|_H^2, \tag{6.68}$$

where $f \in \overline{H}$ represents a vector in $\overline{\mathcal{H}}$ in terms of the L^2 function space $L^2(-\infty, \infty, H)$, and $f_s \in H$, the L^2 function space representing $\mathcal{H}_s$ for any s. The Sinai theorem furthermore asserts that there are representations for which the action of the full evolution group $U(\tau)$ on $L^2(-\infty, \infty, H)$ is translation by τ units. Given $D_{\pm}$ (the L^2 spaces representing $\mathcal{D}_{\pm}$), there is such a representation, called the *incoming translation representation*, for which functions in D_- have support in $L^2(-\infty, 0, H)$, and another called the *outgoing translation representation*, for which functions in D_+ have support in $L^2(0, \infty, H)$. It is clear that s has the interpretation of an observable; it is conjugate to the "energy" σ defined by Lax and Phillips in the so-called energy representation defined by Fourier transform, as we shall see below.

In the discussion of the bound state problem, we considered the system in the total rest frame, for which this component P^0 is the rest energy of the system (its "rest mass"). The invariant $P^{\mu} P_{\mu}$ is the so-called Mandelstam variable corresponding to the mass of the decaying system (or the total mass of the final state of a scattering system in the s-channel).[6] We shall find that the resonance described by the Lax Phillips theory corresponds to a pole in the S-matrix as a function of this variable in the complex plane, as consistent with the usual interpretation of the mass of a resonance. We may therefore identify the foliation variable s with the time T of the center of mass of the decaying system (the fourth component of the center of mass position X^{μ} defined in Chapter 5, a variable covariantly dual to the total center of mass energy, the fourth component of the conserved total energy momentum P^{μ}). The "anti-atom", or anti-two body state, would be associated with

---

[6]In analytic S-matrix theory [Chew (1966), Eden (1967)], the "physical" or direct channel, where the total energy momentum of the incoming (or outgoing) particles is given by, for example, $P^{\mu} = p_1^{\mu} + p_2^{\mu} = p'_1{}^{\mu} + p'_2{}^{\mu}$, is called the s channel. By crossing symmetry, assuming analyticity of the S matrix, one can consider the related process $p_1 + \bar{p}'_1$ in what is called the t channel, or $p_1 + \bar{p}'_2$ in what is called the u channel. Although these substitutions can be made directly in the S-matrix, the corresponding kinematical quantities are non-physical until appropriate analytic continuations are made.

the T-reversed motion (corresponding to time reversal of the entire system).

Lax and Phillips (1967) show that there are unitary operators $W_\pm$, called wave operators, which map elements in $\overline{\mathcal{H}}$, respectively, to these representations. They define an S-matrix,

$$S = W_+ W_-^{-1} \tag{6.69}$$

which connects these representations; it is unitary, commutes with translations, and maps $L^2(-\infty, 0)$ into itself. The singularities of this S-matrix, in what is defined as the *spectral representation*, correspond to the spectrum of the generator of the exact semigroup characterizing the evolution of the unstable system.

With the assumptions stated above on the properties of the subspaces $\mathcal{D}_+$ and $\mathcal{D}_-$, Lax and Phillips [Lax (1967)] gave a simple proof (similar to that given here in (6.72)) that the family of operators

$$Z(\tau) \equiv P_+ U(\tau) P_- \qquad (\tau \geq 0), \tag{6.70}$$

where $P_\pm$ are projections into the orthogonal complements of $\mathcal{D}_\pm$, respectively, is a contractive, continuous, semigroup. This operator annihilates vectors in $\mathcal{D}_\pm$ and carries the space

$$\mathcal{K} = \overline{\mathcal{H}} \ominus \mathcal{D}_+ \ominus \mathcal{D}_- \tag{6.71}$$

into itself, with norm tending to zero for every element in $\mathcal{K}$.

We see from this construction that the outgoing subspace D_+ is defined, in the outgoing representation, in terms of support properties (this is also true for the incoming subspace in the incoming representation). One can then easily understand that the fundamental difference between Lax–Phillips theory and the standard quantum theory lies in this property; the projection operators are associated with *subspaces defined by time*. The subspace defining the unstable system in the standard theory is usually defined as the eigenstate of an unperturbed Hamiltonian, and cannot be associated with an interval on time. The subspaces of the Lax-Phillips theory are associated with intervals (*e.g.*, the positive and negative half-lines in the outgoing and incoming free representations). To see this, we remark that the operator $P_+U(\tau)$ is a semigroup. One has the relation

$$P_+U(\tau_1)P_+U(\tau_2) = P_+U(\tau_1)[1 - (1 - P_+)]U(\tau_2)$$
$$= P_+U(\tau_1)U(\tau_2) = P_+U(\tau_1 + \tau_2); \quad (6.72)$$

which follows from the fact that $U(\tau_1)$ leaves the subspace D_+ invariant.

We now show that the generator of this semigroup is symmetric but not self-adjoint, and it is therefore not a group. In the outgoing translation representation,

$$(P_+U(\tau)f)(s) = \theta(-s)f(s - \tau), \quad (6.73)$$

and therefore

$$(P_+Kf)(s) = i\theta(-s)\frac{\partial f}{\partial s}(s - \tau)|_{\tau \to 0_+}, \quad (6.74)$$

where $f(s)$ is a vector-valued function, and K is the self-adjoint generator associated with $U(\tau)$. If we then compute the scalar product of the vector given in (6.10) with a vector g, we find that

$$\int_{-\infty}^{\infty} ds\, g^*(s)(P_+Kf)(s) = i\delta(s)g^*(0)f(0) + \int_{-\infty}^{\infty} ds(P_+Kg)^*(s)f(s).$$
$$(6.75)$$

The generator is therefore not self-adjoint. It is through this mechanism that the Lax–Phillips theory provides a description that has the semigroup property for the evolution of an unstable system [Horwitz (1973)]. It has, in fact, a family of complex eigenvalues $\{\mu\}$ in the upper half-plane; the eigenfunctions are

$$f_\mu(s) = \begin{cases} e^{\mu s}n, & s \le 0; \\ 0, & s > 0, \end{cases} \quad (6.76)$$

where n is some vector in the auxiliary space.

The semigroup property of the operator $Z(\tau)$ of (6.70) follows directly from the discussion given above. It clearly vanishes on the subspace D_-, and by the stability of D_+ under $U(\tau)$ for $\tau \ge 0$, it vanishes on D_+ as well. It is therefore non-zero only on the subspace K, and on such vectors, the operator P_- can be omitted; the semigroup property then follows from what we have said above.

If we identify elements in the space $\overline{\mathcal{H}}$ with *physical states*, and identify the subspace $\mathcal{K}$ with the unstable system, we see that the quantum Lax Phillips theory provides a framework for the description of an unstable system which decays according to a semigroup law. We remark that, taking a vector ψ_0 in $\mathcal{K}$, and evolving it under the action of $U(\tau)$, the projection back into the original state is (this follows from (6.70) and the stability of $\mathcal{D}_\pm$ that $Z(\tau) = P_\mathcal{K} U(\tau) P_\mathcal{K}$ as well)

$$\begin{aligned}
A(\tau) &= (\psi_0, U(\tau)\psi_0) \\
&= (\psi_0, P_\mathcal{K} U(\tau) P_\mathcal{K} \psi_0) \\
&= (\psi_0, Z(\tau)\psi_0),
\end{aligned} \tag{6.77}$$

so that the survival amplitude (6.58) of the Lax-Phillips theory, analogous to that of the Wigner–Weisskopf formula (6.58), has the exact exponential behavior. The difference between this result and the corresponding expression for the Wigner–Weisskopf theory can be accounted for by the fact that there are translation representations for $U(\tau)$, and that the definition of the subspace $\mathcal{K}$ is related to the support properties along the foliation axis on which these translations are induced.

Functions in the space $\overline{H}$, representing the elements of $\overline{\mathcal{H}}$, depend on the variable s as well as the variables of the auxiliary space H. The measure space of this Hilbert space of states is one dimension larger than that of a quantum theory represented in the auxiliary space alone (the additional dimension may correspond to the center of mass time of the resonance, as pointed out above, a cyclic variable in, for example, the two body problem treated in Chapter 5). With this identification, we may understand this representation of a state as a *virtual history*. The collection of such histories forms a quantum ensemble; the absolute square of the wave function corresponds to the probability that the system would be found, as a result of measurement, at time s in a particular configuration in the auxiliary space (in the state described by this wave function), *i.e.*, an element of one of the virtual histories [Eisenberg (1997)]. For example, the expectation value of the position variable x at a given s is, in the standard interpretation of the auxiliary space as a space of quantum states,

$$\langle x \rangle_s = \frac{(\psi_s, x\psi_s)}{\|\psi_s\|^2}. \tag{6.78}$$

The full expectation value in the physical Lax-Phillips state, according to (6.68), is then

$$\int ds\, (\psi_s, x\psi_s) = \int ds\, \|\psi_s\|^2 \langle x \rangle_s, \tag{6.79}$$

so we see that $\|\psi_s\|^2$ corresponds to the probability to find a signal which indicates the presence of the system at the time s (in the same way that x is interpreted as a dynamical variable in the quantum theory).

One may ask, in this framework, which results in a precise semigroup behavior for an unstable system, whether such a theory can support as well the description of stable systems or a system which makes a transition following the rule of Wigner and Weisskopf (as, for example, the adiabatic rotation of an atom with spin in an electromagnetic field). It is clear that if $D_\pm$ span the whole space, there is no unstable subspace, and one has a scattering theory without the type of resonances that can be associated with unstable systems.

In the following, we give a procedure [Strauss (2000a)] for the construction of the subspaces $D_\pm$, and for defining the representations which realize the Lax-Phillips structure. In this framework, we shall define the Lax-Phillips S-matrix.

It follows from the existence of the one-parameter unitary group $U(\tau)$ which acts on the Hilbert space $\overline{\mathcal{H}}$ that there is an operator K which is the generator of dynamical evolution of the physical states in $\overline{\mathcal{H}}$; we assume that there exist *wave operators* $\Omega_\pm$ which intertwine this dynamical operator with an unperturbed dynamical operator K_0.

We shall assume that K_0 has only absolutely continuous spectrum in $(-\infty, \infty)$ (we discuss below an example in which these assumptions are are explicitly valid).

We begin the development of the quantum Lax-Phillips theory [Strauss (2000a)] with the construction of the incoming and outgoing translation representations. In this way, we shall construct explicitly the foliations required. The *free spectral representation* of K_0 is defined by

$$_f\langle\sigma\beta|K_0|g\rangle = \sigma \,_f\langle\sigma\beta|g\rangle, \tag{6.80}$$

where $|g\rangle$ is an element of $\overline{\mathcal{H}}$ and β corresponds to the variables (measure space) of the auxiliary space associated to each value of σ, which, with σ (identified with the rest energy of the unstable system, as we have pointed out above), comprise a complete spectral set. The functions $_f\langle\sigma\beta|g\rangle$ may be thought of as a set of functions of the variables β indexed on the variable σ in a continuous sequence of auxiliary Hilbert spaces isomorphic to H.

We now proceed to define the incoming and outgoing subspaces $\mathcal{D}_\pm$. To do this, we define the Fourier transform from representations according to the spectrum σ to the foliation variable s of (6.68), *i.e.*,

$$_f\langle s\beta|g\rangle = \int e^{i\sigma s} \,_f\langle\sigma\beta|g\rangle d\sigma. \tag{6.81}$$

Clearly, K_0 acts as the generator of translations in this representation. We shall say that the set of functions $_f\langle s\beta|g\rangle$ are in the *free translation representation*.

Now consider the sets of functions with support in $L^2(0,\infty)$ and in $L^2(-\infty,0)$, and call these subspaces $D_0^\pm$. The Fourier transform back to the free spectral representation provides the two sets of Hardy class functions [Riesz (1952)]

$$_f\langle\sigma\beta|g_0^\pm\rangle = \int e^{-i\sigma s} \,_f\langle s\beta|g_0^\pm\rangle ds \in H_\pm, \tag{6.82}$$

for $g_0^\pm \in D_0^\pm$.

We may now define the subspaces $\mathcal{D}_\pm$ in the Hilbert space of states $\overline{\mathcal{H}}$. To do this we first map these Hardy class functions in $\overline{H}$ to $\overline{\mathcal{H}}$, *i.e.*, we define the subspaces $\mathcal{D}_0^\pm$ by

$$\int \sum_\beta |\sigma\beta\rangle_f {}_f\langle\sigma\beta|g_0^\pm\rangle d\sigma \in \mathcal{D}_0^\pm. \tag{6.83}$$

We shall assume that there are wave operators which intertwine K_0 with the full evolution K, i.e., that the limits

$$\lim_{\tau \to \mp\infty} e^{iK\tau} e^{-iK_0\tau} = \Omega_\pm \qquad (6.84)$$

exist on a dense set in $\overline{\mathcal{H}}$. We emphasize that the operator K generates evolution of the entire virtual history, *i.e.*, of elements in $\overline{\mathcal{H}}$, and that these wave operators are defined in this larger space. These operators are *not*, in general, the usual wave (intertwining) operators for the perturbed and unperturbed Hamiltonians that act in the auxiliary space. The conditions for their existence are, however, closely related to those of the usual wave operators. For the existence of the limit, it is sufficient that for $\tau \to \pm\infty$, $\|Ve^{-iK_0\tau}\phi\| \to 0$ for a dense set in $\overline{\mathcal{H}}$. As for the usual scattering theory, it is possible to construct examples for which the wave operator exists if the potential falls off sufficiently rapidly, as discussed above in connection with the standard relativistic scattering theory.

The construction of $\mathcal{D}_\pm$ is then completed with the help of the wave operators. We define these subspaces by

$$\begin{aligned}
\mathcal{D}_+ &= \Omega_+ \mathcal{D}_0^+ \\
\mathcal{D}_- &= \Omega_- \mathcal{D}_0^-.
\end{aligned} \qquad (6.85)$$

We remark that these subspaces are not produced by the same unitary map. This procedure is necessary to realize the Lax-Phillips structure non-trivially; if a single unitary map were used, then there would exist a transformation into the space of functions on $L^2(-\infty, \infty, H)$ which has the property that all functions with support on the positive half-line represent elements of $\mathcal{D}_+$, and all functions with support on the negative half-line represent elements of $\mathcal{D}_-$ in the same representation; the resulting Lax-Phillips S-matrix would then be trivial. The requirement that $\mathcal{D}_+$ and $\mathcal{D}_-$ be orthogonal is not an immediate consequence of our construction; as we shall see, this result is associated with the analyticity of the operator which corresponds to the Lax-Phillips S-matrix.

In the following, we construct the Lax-Phillips S-matrix and the Lax-Phillips wave operators.

The wave operators defined by (6.84) intertwine K and K_0, i.e.,

$$K\Omega_\pm = \Omega_\pm K_0; \qquad (6.86)$$

we may therefore construct the outgoing (incoming) spectral representations from the free spectral representation. Since

$$K\Omega_\pm |\sigma\beta\rangle_f = \Omega_\pm K_0 |\sigma\beta\rangle_f$$

$$= \sigma\Omega_\pm |\sigma\beta\rangle_f, \qquad (6.87)$$

we may identify

$$|\sigma\beta\rangle_{\substack{out \\ in}} = \Omega_\pm |\sigma\beta\rangle_f. \qquad (6.88)$$

The Lax–Phillips S-matrix is defined as the operator on $\overline{H}$ which carries the incoming to outgoing translation representations of the evolution operator K. Suppose g is an element of $\overline{\mathcal{H}}$; its incoming spectral representation, according to (6.82), is

$$_{in}\langle\sigma\beta|g\rangle = {}_f\langle\sigma\beta|\Omega_-^{-1}g\rangle. \qquad (6.89)$$

Let us now act on this function with the Lax-Phillips S-matrix in the free spectral representation, and require the result to be the *outgoing* representer of g:

$$_{out}\langle\sigma\beta|g\rangle = {}_f\langle\sigma\beta|\Omega_+^{-1}g\rangle$$

$$= \int d\sigma' \sum_{\beta'} {}_f\langle\sigma\beta|\mathbf{S}|\sigma'\beta'\rangle_f \; {}_f\langle\sigma'\beta'|\Omega_-^{-1}g\rangle \quad (6.90)$$

where $\mathbf{S}$ is the Lax–Phillips S-operator (defined on $\overline{\mathcal{H}}$). Transforming the kernel to the free translation representation with the help of (6.81), *i.e.*,

$$_f\langle s\beta|\mathbf{S}|s'\beta'\rangle_f = \frac{1}{(2\pi)^2} \int d\sigma d\sigma' \, e^{i\sigma s} e^{-i\sigma' s'} {}_f\langle\sigma\beta|\mathbf{S}|\sigma'\beta'\rangle_f, \quad (6.91)$$

we see that the relation (6.90) becomes, after using Fourier transform in a similar way to transform the *in* and *out* spectral representations to the corresponding *in* and *out* translation representations,

$$_{out}\langle s\beta|g\rangle = {_f}\langle s\beta|\Omega_+^{-1}g\rangle = \int ds' \sum_{\beta'} {_f}\langle s\beta|\mathbf{S}|s'\beta'\rangle_f {_f}\langle s'\beta'|\Omega_-^{-1}g\rangle$$

$$= \int ds' \sum_{\beta'} {_f}\langle s\beta|\mathbf{S}|s'\beta'\rangle_{fin}\langle s'\beta'|g\rangle. \quad (6.92)$$

Hence the Lax–Phillips S-matrix is given by

$$S = \{{_f}\langle s\beta|\mathbf{S}|s'\beta'\rangle_f\}, \quad (6.93)$$

in free translation representation. It follows from the intertwining property (6.86) that

$$_f\langle\sigma\beta|\mathbf{S}|\sigma'\beta'\rangle_f = \delta(\sigma - \sigma')S^{\beta\beta'}(\sigma), \quad (6.94)$$

This result can be expressed in terms of operators on $\overline{\mathcal{H}}$. Let

$$w_-^{-1} = \{{_f}\langle s\beta|\Omega_-^{-1}\} \quad (6.95)$$

be a map from $\overline{\mathcal{H}}$ to $\overline{H}$ in the incoming translation representation, and, similarly,

$$w_+^{-1} = \{{_f}\langle s\beta|\Omega_+^{-1}\} \quad (6.96)$$

a map from $\overline{\mathcal{H}}$ to $\overline{H}$ in the outgoing translation representation. It then follows from (6.92) that

$$S = w_+^{-1}w_-, \quad (6.97)$$

is a kernel on the free translation representation. This kernel is understood to operate on the representer of a vector g in the incoming representation and map it to the representer in the outgoing representation (see [Horwitz (1947)] for a study of pointwise models corresponding to the nonrelativistic limit of the theory described above, for which the generator acts pointwise on the foliation axis).

6.9 Summary

In this chapter, we discussed the general formulation of scattering theory in the framework of the covariant SHP theory, for which dynamical evolution is governed by a universal invariant parameter τ.

For a system of two or more particles, where the motion of each particle is governed by a parameter of evolution, to maintain a dynamical correlation beween the particles, we have argued that it is essential that the parameter of evolution be *universal*. In this sense, the theory we have discussed here (SHP) is much beyond the original discussion of relativistic mechanics, both classical and quantum, presented by Stueckelberg.

Since the Hamiltonian structure of the relativistic dynamics discussed here is closely parallel to that of the non-relativistic theory, one can define, in the same way, asymptotic (for $\tau \to \pm\infty$) states and an S-matrix for quantum scattering.

As for the non-relativistic theory, the description of resonances, relatively long-lived semi-stable states than can occur during a scattering process, has a long history in particle physics. Such relatively long-lived states appear in many cases to have a identities that do not depend on how they are generated, and as such, are assigned names and properties that qualify them as *particles*. Their decay characteristics generally have the *semigroup* property $Z(\tau_1)Z(\tau_2 = Z(\tau_1 + \tau_2)$. Identifying them with subspaces of the usual Hilbert space, with unitary evolution, however, does not have this observed semi-group property. The theory of Lax and Phillips (1967) discussed here does provide a framework for achieving this experimentally observed property, and is therefore of fundamental importance in relativstic dynamics.

Chapter 7

Gauge Fields for Relativistic Mechanics

In this chapter we discuss the general formulation of gauge fields in the quantum theory, both abelian and nonabelian. A generalization of the elementary Stueckelberg diagram (Fig. 1 of Chapter 2), demonstrating a "classical" picture of pair annihilation and creation, provides a similar picture of a process involving two or more vertices (diagrams of this type appear in Feynman's paper in 1949 [Feynman (1949)] with sharp instantaneous vertices). A single vertex, as it was in Stueckelberg's original diagram, in the presence of a nonabelian gauge field, can induce a transition to an antiparticle with different identity. Such transitions can result in "flavor oscillations", such as in the simple case of neutrino oscillations, transitions between electron and muon neutrinos. We shalll discuss this situation briefly here. The construction of the Lorentz force will also be discussed.

In his original paper Stueckelberg [Stueckelberg (1941)] introduced the electromagnetic vector gauge fields, as we shall explain below, as compensation fields for the derivatives on the wave functions.

For a Hamiltonian of the form (2.45), with $V = 0$, $i.e.$,

$$K = \frac{p^\mu p_\mu}{2M},$$ (7.1)

the Stueckelberg-Schrödinger equation, where p_μ is represented by $-\partial/\partial x^\mu$, is

$$i\frac{\partial}{\partial \tau}\psi_\tau(x) = K\psi_\tau(x).$$ (7.2)

One must introduce so-called compensation fields to retain the form of the equation when the wave function is modified by a (differentiable) phase function at every point [Yang (1954)]. Thus, for

$$\psi(x)' = e^{ie\Lambda(x)}\psi(x), \tag{7.3}$$

the relation

$$(p^\mu - eA^\mu(x)')\psi(x)' = e^{ie\Lambda}(p^\mu - eA^\mu(x))\psi(x), \tag{7.4}$$

where e, the charge on the particle, is satisfied if

$$A^\mu(x)' = A^\mu(x) + \partial^\mu\Lambda. \tag{7.5}$$

One sees that the gauge transformation induced on the compensation field is of the same form as the gauge transformations of the Maxwell potentials, preserving the Gauss and Stokes laws [Jackson (1974)] and therefore this procedure may be thought of as an underlying theory for electromagnetism [Wu (1975)].

Stueckelberg (1941) noted that he was unable to explain the diagram of Fig. 1 with this form of the electromagnetic interaction. The reason is that the canonical velocity is

$$\dot{x}^\mu = \frac{p^\mu - eA^\mu}{M}, \tag{7.6}$$

so that

$$\dot{x}^\mu \dot{x}_\mu = -\left(\frac{ds}{d\tau}\right)^2 = \frac{(p^\mu - eA^\mu)(p_\mu - eA^\mu)}{M^2}. \tag{7.7}$$

This expression is proportional to the conserved Hamiltonian, so that the proper time cannot go through zero. To avoid this difficulty, he added an extra force term in the equations of motion. However, this construction did not take into account the compensation field required for the τ derivative in the Stueckelberg-Schrödinger equation.

For the nonrelativistic Schrödinger equation, the t derivative in the equation requires a compensation field A^0 (in addition to the **A** fields compensating for the action of the derivatives $\mathbf{p} = -i\frac{\partial}{\partial\mathbf{x}}$, thus providing the full set of Maxwell fields. Taking this requirement into account in the Stueckelberg-Schrödinger equation, we arrive at a *five dimensional generalization* of the Maxwell theory [Saad (1989); see also Wesson(2006)]. We furthermore recognize that since the phase may depend on τ, the compensation fields, which we shall denote by a_μ, a_5, must also depend on τ. We shall see that under integration over τ, *i.e.*, the zero mode, the fields a_μ reduce to the usual Maxwell fields satisfying the usual Maxwell equations, and the a_5 field decouples. The more general theory therefore properly contains the Maxwell theory.

We first remark that a_5 and a_μ must tranform under a gauge change according to

$$a_5(x,\tau)' = a_5(x,\tau) + \frac{\partial \Lambda}{\partial \tau}$$
$$a_\mu(x,\tau)' = a_\mu(x,\tau) + \frac{\partial \Lambda}{\partial x^\mu}, \tag{7.8}$$

or, with $\alpha = (0,1,2,3,5)$, and $x^5 \equiv \tau$,

$$a_\alpha(x,\tau) = a_\alpha(x,\tau) + \frac{\partial \Lambda}{\partial x^\alpha}. \tag{7.9}$$

The Stueckelberg-Schrödinger evolution operator in the presence of this $5D$ gauge field must therefore have, minimally, the form

$$i\frac{\partial \psi_\tau(x)}{\partial \tau} = \{\frac{(p^\mu - e'a^\mu)(p_\mu - e'a_\mu)}{2M} - e'a^5(x)\}\psi_\tau(x), \tag{7.10}$$

where e' is related to the Maxwell elementary charge e, as we shall see, by a dimensional scale factor.

One may extract from (7.10) the form for the corresponding classical Hamiltonian,

$$K = \frac{(p^\mu - e'a^\mu)(p_\mu - e'a_\mu)}{2M} - e'a^5(x). \tag{7.11}$$

In this form, the Stueckelberg trajectory drawn in Fig. 1 is, in priniple, realizable. If $-e'a^5$ reaches a value equal to K, these terms can cancel; at these points the proper time interval can pass through zero, and the semiclassical picture of pair annihilation becomes consistent in a simple way.

It follows from the form of (7.9) that the quantities (we use $\partial_\alpha \equiv \frac{\partial}{\partial x^\alpha}$)

$$f_{\alpha\beta}(x,\tau) = \partial_\alpha a_\beta - \partial_\beta a_\alpha \qquad (7.12)$$

are gauge invariant, and may be considered, in analogy to the Maxwell case, and consistently with their interaction, as we shall see below, with the conserved currents, as *field strengths*.

7.1 Transformation Properties and Currents

To consider these quantities as *tensors* requires an additional, very strong assumption, *i.e.*, that the five variables $\{x^\mu, x^5\} \equiv \{x^\alpha\}$, where $x^5 \equiv \tau$ transform together under some group such as $O(3,2)$ or $O(4,1)$. For the latter, the explicit invariance which is evident in the homogeneous equations, that of the Lorentz group, has significant experimental evidence to justify such an assumption; at the present time there is no strong evidence for such a larger symmetry as $O(3,2)$ or $O(4,1)$. We therefore do not assume, *a priori*, the full symmetry under $O(3,2)$ or $O(4,1)$. It is sufficient for our purposes to achieve manifest Lorentz covariance (and Poincaré symmetry for the equations of motion). We now turn to a discussion of the *current* associated with the gauge invariant Stueckelberg equation.

The nonrelativistic fully gauge invariant Schrödinger equation is

$$i\frac{\partial}{\partial t}\psi_t(\mathbf{x} = \frac{(\mathbf{p} - e\mathbf{A}(\mathbf{x},t))^2}{2M}\psi_t(\mathbf{x}) - eA_0; \qquad (7.13)$$

the current $\mathbf{J}$ and the charge density $J^0 \equiv \rho$ satisfy the conservation law

$$\nabla \cdot \mathbf{J} + \frac{\partial \rho}{\partial t} = 0, \qquad (7.14)$$

where

$$\mathbf{J} = \frac{ie}{2M}[\psi^*(\nabla - ie\mathbf{A})\psi - \psi(\nabla + ie\mathbf{A})\psi^*], \qquad (7.15)$$

and

$$J^0 = \rho = e\psi^*\psi. \tag{7.16}$$

The inhomogeneous Maxwell field equations, written formally in terms of four-vector indices, are *e.g.*[Jackson (1974), Landau(1951)]

$$\partial_\nu F^{\mu\nu} = eJ^\mu; \tag{7.17}$$

they may be obtained from a Lagrangian providing the Schrödinger equation as a field equation, with a gauge invariant term proportional to $F_{\mu\nu}F^{\mu\nu}$, as we shall describe below in our discussion of the $5D$ fields).

For the relativistic case, Jackson (1974) has shown how one can construct a covariant four vector current from a sequence of elementary charged *events* in spacetime, which we shall refer to again below. It is, however, important to note that the *homogeneous* equations corresponding to (7.17), *i.e.*, for $J^\mu = 0$, reflect the Lorentz symmetry, suggesting that such a symmetry may indeed be a symmetry of the world. To realize this symmetry consistently, one must use a form of the quantum theory that gives rise to a covariant four current based, as we see above, on (7.10).

7.2 Field Equations

A simple set of field equations, providing second order derivatives of the potentials, is obtained by considering the Lagrangian density due to the field variables to be of the form $f_{\alpha\beta}f^{\alpha\beta}$, where we leave open for now the question of choosing a signature for raising and lowering the index of the fifth component. Writing a Lagrangian density for which setting the coefficient of the variation of ψ^* equal to sero gives the Stueckelberg-Schrödinger equation,[1] with this additional term for

---

[1]Gottfried [Gottfried (1966)] has pointed out that this procedure is not completely consistent since the Schrödinger wave ψ is not a mechanical quantity; it is, however, consistent for quantum field theory, and provides a convenient procedure to generate field equations for the first quantized theory under discussion here. The method is widely used as a heuristic tool (for example, [Bjorken (1964)]).

the gauge fields of the form

$$\mathcal{L} = \frac{1}{2}\left(i\frac{\partial\psi}{\partial\tau}\psi^* - i\psi\frac{\partial\psi^*}{\partial\tau}\right)$$

$$- \frac{1}{2M}\left[(p^\mu - e'a^\mu)\psi((p_\mu - e'a_\mu)\psi)^*\right]$$

$$+ e'(a^5\psi\psi^*) - \frac{\lambda}{4}f^{\alpha\beta}f_{\alpha\beta}, \qquad (7.18)$$

where λ is, as we discuss below, an arbitrary real dimensional scale factor.

As for the $4D$ Maxwell fields, for which the Lagrangian does not contain $\partial A_0/\partial t$, the Lagrangian does not contain $\partial a_5/\partial\tau$, and therefore a full canonical quantization (as contrasted with path integral approachs such as Fadeev-Popov [Fadeev (1967)]), which requires identification of a canonical momentum for the fields as the derivative of the Lagrangian density with respect to the time derivative of the field, is not easily accessible. There are effective methods, for example, of Henneaux and Teitelboim [Henneaux (1992)] and Haller [Haller (1972)] for dealing with this problem; these methods have been applied by Horwitz and Shnerb [Shnerb (1993)] (see also [Horwitz 2015]) to carry out the canonical quantization of the $5D$ fields. We just remark here that the three photon polarization states (in dimensionality the number of field components minus the two constraints due to Gauss's law and a gauge condition) may fall under the $O(2,1)$ or $O(3)$ symmetry groups; black body radiation appears to point to the $O(2,1)$. Photons in an induced representation of the same type as discussed in Chapter 4, with all photons in the cavity carrying the same $O(2)$ at corresponding points on the orbit would satisfy the requirements of black body radiation (see Chapter 8).

The variation of the potentials a^α in (7.18) provides the field equations

$$\lambda\partial^\alpha f_{\beta\alpha} = j_\beta \qquad (7.19)$$

where

$$j_\mu = \frac{ie'}{2M}\{(\partial_\mu - ie'a_\mu)\psi\psi^* - \psi((\partial_\mu - ie'a_\mu)\psi)^*\}, \qquad (7.20)$$

and

$$j_5 = e'\psi\psi^* \equiv \rho_5. \tag{7.21}$$

As for the nonrelativistic gauge theory based on the Schrödinger equation, there is no coordinate transformation which can induce a linear combination of j_μ and j_5, and therefore these equations cannot be covariant under $O(4,1)$ or $O(3,2)$, although the homogeneous form of (7.19) for $j^\alpha = 0$ does admit such a higher symmetry.[2]

Furthermore, the current j^α satisfied, as follows from the Stueckelberg-Schrödinger equation, the conservation law

$$\partial_\alpha j^\alpha = 0 \tag{7.22}$$

In general, then, the current j^μ, cannot be the conserved Maxwell current [Saad (1989)]; see also [Stueckelberg (1941)]. Writing Eq. (7.22) in the form

$$\partial_\mu j^\mu + \frac{\partial \rho}{\partial \tau} = 0 \tag{7.23}$$

suggests taking the integral over all τ [Stueckelerg (1941)]. If $\rho_\tau(x) \to 0$ for $\tau \to \pm\infty$, that is, that the expectation of the occurrence of events in a finite region of x^μ vanishes for large values of the evolution parameter (the physical system evolves out of the laboratory), then the second term vanishes under this integration, and one finds that

$$\partial_\mu J^\mu = 0, \tag{7.24}$$

---

[2]If such a higher symmetry, such as $O(3,2)$ or $O(4,1)$ were to be found as a general property of particle kinematics, such as Lorentz covariance, in the framework of our present experimental knowledge, then a generalization of the Stueckelberg theory could be written with five momenta transforming under this group. The corresponding gauge fields would then be one dimension higher, to take into account the evolution of the system, and the resulting homogeneous field equations would appear to be $O(4,2)$, $O(3,3)$. or $O(5,1)$ invariant. The corresponding theory of spin, as worked out in the previous chapter, would then rest on the method of Wigner applied to the stability group of a five-vector. In this chapter, we shall restrict our analysis to systems which are manifestly covariant on the level of the Lorentz group.

where

$$J^\mu(x) = \int_{-\infty}^{+\infty} d\tau\, j_\tau^\mu(x) \qquad (7.25)$$

can be identified as the Maxwell current (this procedure has been called "concatenation"[Horwitz (1982)]).

In his book on electrodynamics, Jackson (1974) provides a construction of a covariant current by starting with an elementary current element $e\dot{x}^\mu\delta^4(x - x(s))$, where s is considered to be some parameter along the worldline $x^\mu(s)$ of the moving charged event, say, the proper time. He then asserts that

$$J^\mu(x) = e\int ds\, \dot{x}^\mu \delta^4(x - x(s)) \qquad (7.26)$$

is conserved by noting that

$$\begin{aligned}
\partial_\mu J^\mu(x) &= e\int ds\, \dot{x}^\mu \partial_\mu \delta^4(x - x(s)) \\
&= -e\int \frac{d}{ds}\delta^4(x - x(s)),
\end{aligned} \qquad (7.27)$$

which vanishes if the worldline moves out of the range of the laboratory as $s \to \pm\infty$. The transition from (7.25) to (7.26) follows from noting the identity

$$-\frac{d}{ds}\delta^4(x - x(s)) = \dot{x}^\mu \partial_\mu \delta^4(x - x(s)); \qquad (7.28)$$

This is, however, precisely the conservation law (2.21) for the case $\rho(x) = \delta^4(x - x')$ for a charged event at the point x'. It follows from Jackson's construction that what is considered a "particle", in electromagnetism, *i.e.* an object which satisfies a law of conserved current and charge (or probability density), corresponds to at least a large segment of a worldline [Land (1998)], an essentially nonlocal object in the Minkowski space. The nonrelativistic Schrödinger equation has a locally defined conserved current; the bilinear density $\psi_{NR}(\mathbf{x})^*\psi_{NR}(\mathbf{x})$ contains the product of wave functions of precisely equal mass.

As we have seen, however, in Chapter 2, *e.g.* (2.27), the Stueck-elberg wave function for a free particle evolves according to

$$\psi_\tau(x) = U(\tau)\psi(x) = \frac{1}{(2\pi)^2} \int d^4p\, e^{-i\frac{p^\mu p_\mu}{2M}\tau} e^{-ip^\mu x_\mu}\psi(p); \qquad (7.29)$$

since $p^\mu p_\mu = -m^2$, the variable corresponding to the measured mass, the τ integration of the bilinear has the effect of reducing this form to an integral over a bilinear diagonal in the mass. Thus, the τ integration is associated with the retrieval of "particle" properties, as in our discussion of the Newton–Wigner problem in Chapter 2.

Turning to the field equations (7.19), we see that an integral over τ, assuming the asymptotic vanishing of the $f_{\mu5}$ field for $\tau \to \pm\infty$, results (for the μ components) in

$$\lambda\partial^\nu \int d\tau f_{\mu\nu}(x,\tau) = \int d\tau j_\mu(x,\tau); \qquad (7.30)$$

the right hand side corresponds, as we have argued, to the conserved current of Maxwell, so that we may identify, from (7.12),

$$\int d\tau a_\mu(x,\tau) = A_\mu(x), \qquad (7.31)$$

i.e., the Maxwell τ-independent field. Thus the Maxwell field emerges as the zero mode of the fields $a_\mu(x,\tau)$, which we have called the "pre-Maxwell" fields [Saad (1989)]. Due to the linearity of the field equations, the integral over the field equations (7.19) reduce precisely, as we have seen, to the standard Maxwell form.

The physical situation that we have described here corresponds to the emergence of the Maxwell fields from detection apparatus that intrinsically integrates over τ. It would appear that there is, according to this theory, a high frequency modulation of the Maxwell field that is not easily observable in apparatus available in laboratories at the present time. There has been some indirect evidence, in connection with the self-interaction problem, for the existence of the classical $5D$ fields in connection with an extensive investigation of the self-interaction problem [Aharonovich (2011)]. Furthermore, the fifth field, as we have pointed out above, can be responsible for the transition represented in Stueckelberg's diagram

Fig. 4; it also plays an essential role in the neutrino oscillation model
that we shall describe below.

(7.31) implies that the dimensionality of the pre-Maxwell fields;
since the Maxwell fields A have dimensionality L^{-1}, must be L^{-2}.
Thus the charge that we have called e' must have dimension L (p^μ
has dimension L^{-1}). The gauge invariant field strengths then have
dimension L^{-3}. The quadratic contribution of the field strengths to
the Lagrangian, $f^{\alpha\beta}f_{\alpha\beta}$ then has dimension L^{-6}. Since the action is
an integral of the Lagrangian density over $d\tau d^4x$, of dimension L^5,
the quadratic field strength terms must have a dimensional factor λ.
The current in the resulting field equations contains the factor e',
and the derivatives of the field strength on the right emerge with a
factor λ; thus we can identify

$$e = e'/\lambda \qquad (7.32)$$

with e the dimensionless Maxwell charge.

Assuming the analog of the Lorentz gauge for the five dimensional
fields (4.11),

$$\partial^\alpha a_\alpha = 0, \qquad (7.33)$$

the field equations (7.19) become

$$(-\partial_\tau^2 + \partial_t^2 - \nabla^2)a_\beta = j_\beta/\lambda, \qquad (7.34)$$

where we have taken the $O(4,1)$ signature for the fifth variable τ.
Representing $a_\beta(x,\tau)$ in terms of its Fourier transform $a_\beta(x,s)$, with

$$a_\beta(x,\tau) = \int ds e^{-is\tau} a_\beta(x,s), \qquad (7.35)$$

one obtains

$$(s^2 + \partial_t^2 - \nabla^2)a_\beta(x,s) = j_\beta(x,s)/\lambda, \qquad (7.36)$$

providing a relation between the off-shell mass spectrum of the a_β
field and the quantum mechanical current source. As we have pointed
out earlier, the solutions of wave equations with a definite mass m
have, according to Newton and Wigner [Newton (1949)], nonlocality
of the order of $1/m$; thus (for application of their arguments, thinking
of the field as the wave function of a quantum of the field) the

massless particle would have a very large support. There is some difficulty in imagining the emission of a photon from an atom of the size 10^{-8} cm which instantaneously has infinite support. However, if the photon being emitted is far off shell, and has an effective mass s, as in the equations above, which is fairly large, the particle being emitted can have very small spatial support, undergoing a relaxation process asymptotically to a particle with very small, essentially zero, mass.

A similar argument can be applied to the photoelectric effect; the energy $\hbar\omega$ associatated with a photon of frequency ω is absorbed by a metal plate, and an electron emitted with exactly this energy (minus the work function to free the electron). The contraction of the energy of a highly nonlocalized radiation field into the very small region occupied by the electron is often attributed to "collapse of the wave function", but this statement does not account for the physical mechanism ("collapse" mechanisms require the construction of a model [Hughston (1996)], see also [Silman (2008)]). In this process, again one may think of the photon going far off mass shell to be able to be absorbed locally.

It has often been argued, moreover, that an experimental bound on the photon mass is provided by gauge invariance. This argument would, of course, provide a bound if the mass term in the field equations had some given constant value; then the shift of the vector potential by a gradient term, even if the gauge function satisfied a homogeneous d'Alembert type equation, would leave an extra term in the equation that would not vanish. However, as we have seen, the field equation contains a second derivative with respect to τ, and if the gauge function has a vanishing $5D$ d'Alembertian of $O(4, 1)$ or $O(3, 2)$ type, gauge invariance would be maintained.

We finally remark that the theory of Glashow, Weinberg and Salam (1970) for the description of weak and electromagnetic interactions, such as involved in the decay of a neutron to proton, neutrino and electron, makes use of *non-Abelian* gauge fields, *i.e.* fields with components that do not commute with each other. Such a theory can also be formulated with a Hamiltonian of the form we have used here. The Stueckelberg process of Fig. 4, however can

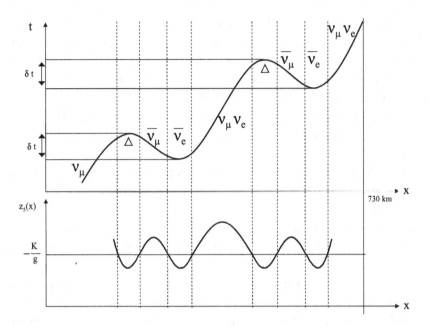

Fig. 4.   This figure shows the "pull-back" in time due to neutrino oscillations arriving at the Gran Sasso 730 km from their source at CERN; the lower part shows the concomitant variations in the fifth field. [Horwitz (2015)] (Courtesy Springer, Dordrecht).

carry a particle to a different identity after the vertex, for example, the neutrino associated with neutrino decay or the μ meson decay to electron and μ neutrino. A sequence of such vertices can account for what is called "neutrino oscillations". The segments going backward in time t can result in an arrival time for a travelling beam of neutrinos which is reduced, giving an impression of faster than ligh speed transmission.[3]

---

[3]During transmission of an electron, its self interaction with the electromagnetic field would have a similar effect [Feynman (1949)], with arrival time less than distance divided by velocity, but since its speed is generally far below light speed, a small early arrival would be difficult to measure. For the neutrino case, there is some experimental indication, but not definitive. [Grand Sasso (2011)], as discussed in more detail in [Horwitz 2015].

Chapter 8

Relativistic Classical and Quantum Statistical Mechanics

In this chapter, we shall discuss the relativistic statistical mechanics of a many body system, for which the points in space time constitute the fundamental entities for which distribution functions must be constructed to achieve a manifestly covariant theory. The counting of events is essentially equivalent to the counting of world lines corresponding to *particles*. Therefore one should expect that, as we indeed find, the statistical mechanics of events is therefore closely related to the theory of statistical mechanics of particles, as developed, for example, in Synge [Synge (1957); see also, de Groot (1980)]. We construct a canonical Gibbs ensemble based on a microcanonical ensemble, enabling us to define a temperature and the basic thermodynamic functions [Horwitz (1981)].

We obtain the Bose-Einstein and Fermi-Dirac statistics from the relativistic kinetic theory with a form analogous to that of the nonrelativistic theory.

In the induced representation in the Fock space for bosons in the quantum theory, we find the correct factor of 2 for the specific heat, even though, in principle, the 5D gauge fields contain, under canonical quantization [Henneaux (1992)], intrinsically three polarization states [Shnerb (1993)].

In this framework, we develop a covariant Boltzmann equation. The usual method of derivation of the second law from the Boltzmann equation [Huang (1967)] shows that the entropy in the relativistic case increases monotonically in τ, but if antiparticles are present, not necessarily in t. Since t is the observed time of events, one may therefore see, in this case, an entropy decrease [Horwitz (2015)] for an isolated system.

8.1 A Potential Model for the Many Body System

We shall consider a model for the many body system, as discussed in Chapter 5, with particle (event) coordinates and momenta $\{q_i^\mu\}$ and $\{p_i^\mu\}$, for $i = 1, 2, \ldots N$, in which the total invariant Hamiltonian is given by

$$K = K_0 + V(q_1, q_2, \ldots, q_N), \tag{8.1}$$

where

$$K_0 = \Sigma_i \frac{p_i{}^\mu p_{i i \mu}}{2M_i}, \tag{8.2}$$

In the following, we treat the center of mass as an additional degree of freedom; it acts as a free particle with momentum $P^\mu = \Sigma_i p_i^\mu$. We furthermore consider here the state for which the center of mass is on-shell, $i.e.$, $P_\mu P^\mu = -Mc^2$.

8.2 The Microcanonical Ensemble

The micronanical ensemble in classical nonrelativistic statistical mechanics is constructed [Huang (1967)] by computing the volume in phase space accessible to a total energy E. The Hamiltonian (8.1) contains, however, the invariants $p_i{}^\mu p_{i\mu} = \mathbf{p}_i^2 - E_i^2/c^2$, which does not bound the phase space (unlike for the nonrelativistic Hamiltonian containing $\mathbf{p}_i{}^2/2M_i$) .

For the nonrelativistic microcanonical ensemble, the phase space integral is usually confined to a box with perfect reflecting walls so that the energy is not perturbed by collisions with the walls. If we bound our spacetime volume in time, the reflecting walls can generate particles running backward in time, $i.e.$, antiparticles. Although such states should exist, for example, in high temperature plasmas, for the sake of simplicity, we shall assume that the system we are describing constitutes an amorphous cloud of events, which is reasonably bounded in space and time, and retains its structure for a sufficient interval of evolution time τ to be able to think of it as a (quasi-)equilibrium state. The integation on q_i will therefore be restricted to this domain.

Since the entire system is translation invariant in time, the center of mass quantity $\Sigma_i E_i$ is a constant of the motion; with this constraint, we can then write the microcanonical ensemble as

$$\Gamma(\kappa, E) = \int d^4 p_1 d^4 p_2 \ldots d^4 p_N d^4 q_1 d^4 q_2 \ldots d^4 q_N \delta(K - \kappa) \delta$$

$$\times \left(\Sigma_i \frac{E_i}{c} - \frac{E}{c} \right), \tag{8.3}$$

where the integrals over $d^4 p_i = \frac{dE_i}{c} d^3 \mathbf{p_i}$ and $d^4 q_i = c dt_i d^3 \mathbf{q_i}$, are bounded by our assumption on the space-time distribution and the requirement that the $\{E_i\}$ are bounded (thus bounding the $\mathbf{p}_i$).

The four momentum integration is also to be understood as allowing for variations of the particle masses $m_i = \sqrt{((\frac{E_i}{c})^2 - \mathbf{p}_i^2)}$ that are not too large, since we are thinking of applications of the theory to systems of particles reasonably close to their Galilean limiting masses. We work in the framework of systems not too far from having nonrelativistic structure. The structure presents, nevertheless, a wide range of generalization, holding potentially the prediction of phenomena not at all of familiar type.

Following the usual procedure [Huang (1987)], we define the microcanonical entropy as

$$S = k_B \ln \Gamma(\kappa, E) \tag{8.4}$$

and the temperature as

$$T^{-1} = \frac{\partial S(\kappa, E)}{\partial E}. \tag{8.5}$$

Taking $c \to \infty$, we obtain the form of the Galilean limit of the relativistic microcanonical ensemble represented by (8.2). The microcanonical ensemble in this limit should contain definite masses for each particle. This requirement can be made precise by requiring that the quantity

$$\varepsilon_i \equiv E_i - M_i c^2$$

$$= c\sqrt{\mathbf{p}_i^2 + m_i^2 c^2} - M_i c^2, \tag{8.6}$$

constituting a change of variables from energy and momentum to momentum and (variable) mass, be bounded in the limit $c \to \infty$. Expanding in powers of $1/c^2$, one finds that

$$\varepsilon_i = c^2 (m_i - M_i) + \mathbf{p}_i^2 / 2M + O(1/c^2). \tag{8.7}$$

Hence in this limit,

$$m_i = M_i + \eta_i, \tag{8.8}$$

where

$$\eta_i = c^2 (m_i - M_i) \tag{8.9}$$

may have any value in $(-\infty, \infty)$, but does not diverge as $c \to \infty$. This residual quantity preserves the relativistic Poisson brackets, and in the quantum case where the limits are controlled by the structure of the wave function, the commutation relations. We shall see that it is this freedom which permits us to obtain the Galilean microcanonical ensemble.[1]

The kinetic terms of the Hamiltonian (8.1) can then be written

$$\Sigma_i \frac{p_i^\mu p_{i\mu}}{2M_i} = \Sigma_i \frac{-E_i^2/c^2 + \mathbf{p}_i^2}{2M_i}$$

$$= \Sigma_i \frac{\mathbf{p}_i^2}{2M_i} - \varepsilon' - \frac{Mc^2}{2} - \Sigma_i \frac{\varepsilon_i^2}{2M_i c^2}, \tag{8.10}$$

where $\varepsilon' = \Sigma_i \varepsilon_i$, and the last term vanishes as $O(1/c^2)$ since the ε_i are bounded.

---

[1]It was shown by Horwitz and Rotbart [Horwitz (1981a)], by examining the scalar product, that the nonrelativistic limit of the relativistic quantum theory is obtained systematically in this way as well.

Since, according to the equations of motion,

$$\frac{d}{d\tau}ct_i = -\frac{\partial K}{\partial(E_i/c)} = -\frac{\partial K_0}{\partial(\varepsilon_i/c)} = c + \frac{\varepsilon_i}{M_i c}, \qquad (8.11)$$

it follows that

$$ct_i = c\tau + \int_0^\tau \frac{\varepsilon_i(\tau')}{M_i c}d\tau' + ct_i(0), \qquad (8.12)$$

so that all events become synchronized in t in the Galilean limit. Hence, V, which depends only on the differences between the $\{t_i\}$ becomes independent of time in the Galilean limit, and we obtain (setting all the $t_i(0) = 0$)

$$K \sim H - \varepsilon' - Mc^2/2, \qquad (8.13)$$

where

$$H = \Sigma_i \frac{\mathbf{p}_i^2}{2M_i} + V(\mathbf{q}_1, \mathbf{q}_2, \dots \mathbf{q}_N), \qquad (8.14)$$

the usual N body nonrelativistic Hamiltonian.

We now turn to the integal measure on the microcanonical ensemble. Noting that

$$\frac{dE_i}{c} = \frac{c^2 m_i dm_i}{\sqrt{\mathbf{p}_i^2 + m_i^2 c^2}} \sim c\,dm_i, \qquad (8.15)$$

we can write the limiting form of the microcanonical ensemble as

$$\Gamma(\kappa, E) \sim \int c\,dm_1 \dots c\,dm_N d^3 p_1 \dots d^3 p_N d^4 q_1 \dots d^4 q_N$$

$$\cdot \delta(H - \varepsilon' - Mc^2/2 - \kappa)\delta(\varepsilon' - \varepsilon) \cdot c, \qquad (8.16)$$

where $\varepsilon = E - Mc^2$, and the integrals are limited by taking the m_i in a small range μ_i around their Galilean limits, and the q_i in the spacetime volume for the ensemble we have referred to earlier. Now, let $dm_N = (1/c)d\eta_N$ and integrate over this variable in (8.16). The linear occurrence of η_N in both δ-function factors allows us to fold the integration, resulting, for $\kappa = -Mc^2/2$, in the standard form of the nonrelativistic microcanonical ensemble

$$\Gamma(\kappa, E) \sim (\Delta mc^2 \Delta t)^{N-1}(c\Delta t) \int d^3p_1 \ldots d^3p_N d^3q_1 \ldots d^3q^N$$

$$\cdot \delta(H(\mathbf{q}_1 \ldots \mathbf{q}_N, \mathbf{p}_1, \ldots, \mathbf{p}_N) - \varepsilon). \tag{8.17}$$

The factors multiplying the integral are not important in the determination of the mean value of any physical observable. We have therefore shown that the relativistic microcanonical ensemble reduces to the usual Galilean microcanonical ensemble in the non-relativistic limit $c \to \infty$. The total energy is identified with $\Sigma_i \varepsilon_i$. We further remark that the result $\kappa = -Mc^2/2$ implies that the factor $\delta(K - \kappa)$ in (8.8) restricts the center of mass motion described by $P^\mu P_\mu$ to mass shell with shift due to K_{rel}, playing the role of an "internal energy".

We now study some of the properties of a system of free particles in the relativistic microcanonical ensemble. From (8.10), with no coordinate dependence in the integral, the integrals over spacetime just yield the total volume of the admissible space, *i.e.*,

$$\Gamma_{free}(\kappa, E) = V^N (cT)^N \int \frac{dE_1}{c} \ldots \frac{dE_N}{c} d^3p_1 \cdots d^3p_N \cdot$$

$$\times \delta \left(\Sigma_i \frac{\mathbf{p}_i^2}{2M_i} - \frac{E_i^2}{2M_i c^2} - \kappa \right)$$

$$\cdot \delta \left(\Sigma_i \frac{E_i}{c} - \frac{E}{c} \right) \tag{8.18}$$

$$= V^N f(\kappa, E).$$

It then follows that, using the usual thermodynamic relations,

$$S = k_B \ln \Gamma = k_B \ln V^N + k_B \ln f(\kappa, E), \tag{8.19}$$

and

$$\frac{1}{T} = \frac{\partial S}{\partial E} = \frac{k_B}{f(\kappa, E)} \frac{\partial f(\kappa, E)}{\partial E}. \tag{8.20}$$

Differentiating (8.19) with respect to V at constant S, we have

$$0 = k_B \frac{N}{V} + k_B \frac{1}{f} \frac{\partial f(\kappa, E)}{\partial E} (-P), \qquad (8.21)$$

where

$$P = -\left(\frac{\partial E}{\partial V}\right)_S. \qquad (8.22)$$

With (8.20), we then find that the result of Jüttner (1911), *i.e.*

$$PV = Nk_B T \qquad (8.23)$$

is valid for the free relativistic gas.

We now study the *ultrarelativistic limit* for the free gas by considering the microcanonical ensemble for $c \to 0$.

With the change of variables

$$\frac{dE_i}{c} = \frac{m_i dm_i c^2}{\sqrt{\mathbf{p}_i^2 + m_i^2 c^2}} \sim c^2 \frac{m_i dm_i}{p_i}, \qquad (8.24)$$

where $p_i = |\mathbf{p}_i|$, and for $\kappa = -\frac{1}{2}Mc^2$, the microcanonical ensemble then reads

$$\Gamma(\kappa, E) \sim (4\pi)^N V^N T^N c^{3N-1} \int m_1 dm_1 \cdots m_N dm_N p_1 dp_1 \cdots p_N dp_N$$

$$\cdot \delta \left(\Sigma_i \frac{m_i^2}{M_i} - M\right) \delta \left(\Sigma_i E_i - E\right). \qquad (8.25)$$

Since $p_i dp_i = (1/c^2) E_i dE_i$, we evaluate this integral in the range $p_i = (0, \infty)$ for which $E_i = (m_i c^2, \infty) \sim (0, \infty)$; taking into account the δ-function constraints, the integral becomes

$$\Gamma(\kappa, E) \sim (4\pi)^N V^N c^{N-1} \int m_1 dm_1 \cdots m_N dm_N \delta \left(\Sigma_i \frac{m_i^2}{M_i} - M\right)$$

$$\cdot \int_0^E E_1 dE_1 \cdots \int_0^{E-E_1-\cdots-E_{N-1}} E_{N-2} dE_{N-2}$$

$$\cdot \int_0^{E-E_1-\cdots-E_{N-2}} E_{N-1} dE_{N-1}$$

$$\times (E - E_1 - E_2 - \cdots - E_{N-1}). \qquad (8.26)$$

By successive differentiation with respect to E, one finds that the integral has the value $E^{2N-1}/(2N-1)!$, so that

$$\Gamma(\kappa, E) \propto E^{2N-1} \sim E^{2N}. \tag{8.27}$$

It then follows from (8.20) that

$$E = 2Nk_BT. \tag{8.28}$$

The same ultrarelativistic limit is obtained, as we shall see below, from the canonical ensemble. For a system with four degrees of freedom (spacetime), and an energy of $\frac{1}{2}k_BT$ for each degree of freedom, this result is consistent with kinetic theory. The canonical ensemble written by Pauli (1921), based on the exponential of $-c\sqrt{\mathbf{p}^2 + m^2c^2}/k_BT$ with measure d^3p/p^0, although computed in his article for limiting cases only for the nonrelativistic (low temperature) limit, yields $3Nk_BT$ for the ultrarelativistic (high temperature) case, as we shall see below,and does not appear to have a simple kinetic theory interpretation.

Evaluating the relativistic microcanonical phase space integral (8.3) for the case of a single particle, one obtains

$$\Gamma(\kappa = -\frac{1}{2}Mc^2, E) = 4\pi cVTM\sqrt{E^2/c^2 - M^2c^2}, \tag{8.29}$$

so that, with (8.20), one obtains

$$k_BT = p\frac{pc}{E/c} = pv, \tag{8.30}$$

in agreement with Pauli's result; as he remarked, there is no direct connection with equipartition since pv is not connected in a simple way with the energy of the system as it is in Galilean mechanics.

8.3   Canonical Ensemble

The canonical ensemble is extracted from the microcanical ensemble by studying a (small) subsystem s of the system which can exchange only heat (kinetic energy) with the remainder, b, of the system. The usual [Huang (1967)] assumption of short range forces, justifying the decomposition

$$K \cong K_s + K_b \qquad (8.31)$$

must be examined carefully since we are dealing with Lorentz invariant potentials which, in the simplst case, may be considered to be functions of Minkowski squared translation invariant combinations of the q_i's. For the case of two body potentials, for example, the quantity $(q_1 - q_2)^\mu (q_1 - q_2)_\mu$ may be small, and limited by the shape of the potential as a function of these invariants, but both $|\mathbf{q}_1 - \mathbf{q}_2|$ and $c(t_1 - t_2)$ may be large, thus putting into question the possibility of separating the two subsystems by some boundary in space and time. In this far region, however, both events are close to their relative light cone. The surfaces of $V = $ const are hyperbolic and asymptotically approach the light cone. A small timelike or spacelike range (in the corresponding invariant measure of distance) implies that the potential is nonvanishing asymptotically only in a thin shell close to the light cone. The existence of a wave operator in quantum scattering theory (Chapter 6), implying the existence of free asymptotic states in the quantum mechanical problem with potentials of this type [Horwitz(1980), Horwitz (1982)] provides some corroborative evidence for (8.31). We shall assume this mechanism for our classical discusson as well.

The constraint structure of both subsystems, the large and the small, must be of the same form as for the entire system. We therefore express the phase space integral for the microcanical ensemble of the whole system in terms of variables referring to s and b as (for $M = M_s + M_b$)

$$\Gamma(E, M) = \int d^4 p_1 \cdots d^4 p_{N_s} d^4 q_1 \cdots d^4_{N_s}$$

$$d^4 p_{N_s+1} \cdots d^4 p_{N_s+N_b}$$

$$\cdot d^4 q_{N_s+1} \cdots d^4_{N_s+N_b} \delta\left(K_s + \frac{1}{2}M_s c^2\right) \delta\left(K_b + \frac{1}{2}M_b c^2\right)$$

$$\cdot c\delta(E_s + E_b - E), \qquad (8.32)$$

where $E_s = \Sigma_{i=1}^{N_s} E_i$, $E_b = \Sigma_{i=N_s+1}^{N_s+N_b} E_i$, and we have changed our notation labelling the phase space volume from $\Gamma(\kappa = -\frac{1}{2}Mc^2, E)$ to $\Gamma(E, M)$ for brevity.

Comparing with our original definition (8.3) of the micocanonical ensemble, we can write (8.32) as

$$
\Gamma(E, M) = \int d^4p_1 \cdots d^4p_{N_s} d^4q_1 \ldots d^4q_{N_s} \delta
$$

$$
\times \left(K_s + \frac{1}{2}M_s c^2 \right) \Gamma_b(E - E_s, M_b)
$$

$$
= \int \frac{dE'}{c} \int d^4p_1 \cdots d^4p_{N_s} d^4q_1 \ldots d^4q_{N_s} \delta
$$

$$
\times \left(K_s + \frac{1}{2}M_s c^2 \right) \delta(E_s - E') \cdot c
$$

$$
\cdot \Gamma_b(E - E', M_b)
$$

$$
= \int \frac{dE'}{c} \Gamma_s(E'.M_s) \Gamma_b(E - E'.M_b). \tag{8.33}
$$

Following the argument given *e.g.* by [Huang (1967)], we assume that there is a maximum in the integrand that dominates the integral at $E' = \bar{E}$ so that

$$
S = k_B \Gamma(E, M) \cong k_B \ln \Gamma_s(\bar{E}, M_s) + k_B \ln \Gamma_b(E - \bar{E}, M_b)
$$
$$
\cong S_s + S_b. \tag{8.34}
$$

The existence of such a maximum value is consistent with the additivity of the entropy. It follows that (we write the equality assuming that the approximation is very good)

$$
\frac{1}{T} = \frac{\partial S_s}{\partial E'}|_{E'=\bar{E}} = \frac{\partial S_b}{\partial E'}|_{E'=E-\bar{E}}, \tag{8.35}
$$

defining T as a parameter of the equilibrium state. It follows that for a (small) value of $\bar{E} = E_s$,

$$
\Gamma_b(E - E_s, M_b) = e^{S_b(E-E_s, M_b)/k_B}
$$
$$
\cong e^{S_b(E, M_b)/k_B} e^{\frac{-E_s}{k_B T}}. \tag{8.36}
$$

Dropping the subscript s, the normalized distribution for the canonical ensemble is then

$$D(q, p) = \delta(K + Mc^2)e^{-\beta E}/Q_N(V^{(4)}, T, M), \qquad (8.37)$$

where $\beta = \frac{1}{k_B T}$ and the *relativistic partition function* is then given by

$$Q_N(V^{(4)}, T, M) = \int \frac{d^{4N}p \, d^{4N}q}{N! h^{4N}} \delta\left(K + \frac{1}{2}Mc^2\right) e^{-\beta E}$$

$$\equiv e^{-\beta A(V^{(4)}, T, M)}, \qquad (8.38)$$

where we have inserted the constant h of dimension momentum times length, which will occur in numerator and denominator of any expectation value, to make the definition of Q_N dimensionless. The integration in (8.38) is taken over the constraint $E = \Sigma E_i$, $q_i \in \sigma_i$(subsystem) and m_i in a small range around the Galilean masses. The formula (8.38) defines the quantity A that, as we shall show below, can be identified with the Helmholtz free energy. Bringing the exponential to the left side, and differentiating with respect to β,

$$\int \frac{d^{4N}p \, d^{4N}q}{N! h^{4N}} \delta\left(K + \frac{1}{2}Mc^2\right) e^{\beta(A-E)} = 1 \qquad (8.39)$$

one finds

$$A = \langle E \rangle - \beta \left(\frac{\partial A}{\partial \beta}\right)_{V^{(4)}} = \langle E \rangle - T \left(\frac{\partial A}{\partial T}\right)_{V^{(4)}}$$

$$\equiv \langle E \rangle - TS, \qquad (8.40)$$

consistent with the interpretation of A as the Hemholtz free energy. We remark that the relation between the relativistic entropy and the relativistic Helmholtz free energy differs from the nonrelativistic case in that the *four-dimensional volume* must be held constant.

Fluctuations in energy may be estimated in the usual way [Huang (1967)] by differentiating

$$0 = \int \frac{d^{4N}p \, d^{4N}q}{N! h^{4N}} \delta\left(K + \frac{1}{2}Mc^2\right) (E - \langle E \rangle)e^{\beta(A-E)} \qquad (8.41)$$

with respect to β at constant four volume $V^{(4)}$ to obtain

$$\langle (E - \langle E \rangle)^2 \rangle = -\frac{\partial \langle E \rangle}{\partial \beta} = k_B T^2 \frac{\partial \langle E \rangle}{\partial T}. \tag{8.42}$$

Assuming that the absorption of heat does not change the time interval Δt of the ensemble, the definition

$$\left(\frac{\partial \langle E \rangle}{\partial T} \right)_{V^{(4)}} = C_V \tag{8.43}$$

coincides with the usual definition of specific heat in the nonrelativistic limit.

The partition function (8.38) can be written as an integral over microcanonical ensembles (8.3) as (absorbing the factors $N! h^{4N}$ into Γ for now)

$$Q_N(V^{(4)}, T, M) = \int \frac{dE}{c} e^{-\beta E} \Gamma \left(-\frac{1}{2} M c^2, E \right). \tag{8.44}$$

Since, in the microcanonical ensemble, $S^{mic} = k_B \ln \Gamma$,

$$Q_N(V^{(4)}, T, M) = \int \frac{dE}{c} e^{\beta(T S^{mic}(E) - E)}. \tag{8.45}$$

The principal contribution to the integral is at the maximum (if there is more than one, there may be phases, as discussed in [Huang (1967)] of the exponent, *i.e.*, where

$$T \frac{\partial S^{mic}(E)}{\partial E} - 1 = 0, \quad \frac{\partial^2 S^{mic}(E)}{\partial E^2} < 0. \tag{8.46}$$

The first of (8.46), coinciding with the definition of temperature in the microcanonical ensemble, implies that the stationary point occurs at $E = \langle E \rangle$, and the second corresponds to

$$\frac{\partial}{\partial E} \frac{1}{T(E)} = -\frac{1}{T^2} \frac{1}{C_V} < 0, \tag{8.47}$$

an expected property for physical systems.

Evaluating the integral (8.45) after expanding to second order, one finds, as in the nonrelativistic theory, that the definitions of entropy for the canonical and microcanonical ensemble differ only in terms of the order $\log \langle N \rangle$. It is important to remember that in this analysis we are discussing the statistical mechanics of *events* rather than *particles*. However, we understand that on the condition that every event is a point along a trajectory in spacetime that corresponds to a particle [Weyl (1952)], in the sense we have explained in Chapter 7, *i.e.*, the object constituting a conserved four vector current is represented in a covariant way as an integral over the world line [Jackson (1974)]. Therefore, the counting of events in covariant statistical mechanics can be understood as a counting of particles. Covariant statistical mechanics should, as we have shown above, have a close relationship, with essentially the same meaning for the thermodynamic potentials, to the statistical mechanics of particles associated with world lines [Synge (1952)].

Following the arguments for the nonrelativistic limit of the microcanonical ensemble given above, one finds that

$$Q_N(V^{(4)}, T, M) \sim \frac{(c\Delta t)^N (c\Delta m)^{N-1}}{N! h^{4N}} \frac{1}{c} e^{-\beta M c^2}$$

$$\int d^3 p_1 \cdots d^3 p_N d^3 q_1 \cdots d^3 q_N e^{-\beta H}, \quad (8.48)$$

The relativistic canonical ensemble therefore reduces to the usual nonrelativistic one in the limit $c \to \infty$ (the factor $e^{-\beta M c^2}$ corresponds to a shift in A (8.40)).

For the relativistic free gas in the canonical ensemble, let us take all the $M_i = M_0$. To compare with the usual formulation, for example, of Jüttner (1911) and Pauli (1921), we shall restrict the range (δm) of masses to be close to the common value M_0. It then follows that

$$Q_N(V^{(4)}, T, M) \cong \frac{c^{2N-2}(c\Delta t)^N (\delta m)^{N-1}}{N! h^{4N}} 2M_0$$

$$\times \left(V \int d^3 p \frac{e^{-\beta c \sqrt{\mathbf{p}^2 + M_0^2 c^2}}}{\sqrt{\mathbf{p}^2 + M_0^2 c^2}} \right)^N. \quad (8.49)$$

The last factor in (8.49) may be compared to Pauli's formula[Pauli (1921)]

$$Q^{Pauli} = \left(V \int d^3 p e^{-\beta c \sqrt{\mathbf{p}^2 + M_0^2 c^2}} \right)^N, \quad (8.50)$$

which differs from our approximate expression (8.49), most significantly at large $\mathbf{p}^2$.

Using the relation

$$\langle E \rangle = -\frac{\partial}{\partial \beta} \ln Q_N, \tag{8.51}$$

the result of the integral [Pauli (1921)] for (8.50) is

$$\langle E \rangle = N k_B T \left\{ 1 - \sigma \frac{i H_2'^{(1)}(i\sigma)}{i H_2^{(1)}} \right\}, \tag{8.52}$$

where $H_2^{(1)}$ is the Hankel function of the second kind, and $\sigma = M_0 c^2 / k_B T$. For σ large (c large or T small, the nonrelativistic limit), one finds [Pauli (1921)]

$$\langle E \rangle \sim \frac{3}{2} N k_B T + N M_0 c^2, \tag{8.53}$$

and for σ small (c small or T large, the relativistic limit),

$$\langle E \rangle \sim 3 N k_b T. \tag{8.54}$$

The first result, obtained explicitly by Pauli shows consistency with the physics of the nonrelativistic limit of his form of the canonical ensemble, but the last result, not reported in Pauli's work, indicates that in the relativistic limit there is no simple interpretation using the rule of $\frac{1}{2} k_B T$ for each degree of freedom. However, for the covariant ensemble of (8.49), using the result

$$\int d^3 p \frac{e^{-\beta c \sqrt{\mathbf{p}^2 + M_0^2 c^2}}}{\sqrt{\mathbf{p}^2 + M_0^2 c^2}} = -2\pi^2 \frac{M_0}{\beta} H_1^{(1)}(i\sigma), \tag{8.55}$$

where $H_1^{(1)}$ is the first type Hankel function, is that for the nonrelativistic limit,

$$\langle E \rangle \sim \frac{3}{2} N k_B T, \tag{8.56}$$

and in the relativistic limit

$$\langle E \rangle \sim 2 N k_B T. \tag{8.57}$$

The latter result is consistent with $\frac{1}{2}k_B T$ for each degree of freedom (there are four, for spacetime). This regime should be valid for $T > 10^{12}$ degrees Kelvin.

8.4 Grand Canonical Ensemble

We shall consider the grand canonical ensemble as composed of a set of subsystems corresponding to canonical ensembles generated by the exchange of both heat energy and particles with the bath. We cannot specify in advance the quantity $M = \Sigma_i M_i$ because the number of particles is indefinite. There are $N!/N_s!(N - N_s)!$ ways of selecting a subsystem of N_s particles, and therefore the canonical partition function for this set may be written

$$Q_N(V^{(4)}, T, M) = \Sigma_{N_s+N_b=N} \int \frac{d^{4N_s}p \, d^{4N_s}q \, d^{4N_b}p \, d^{4N_b}q}{N! h^{4N}}$$

$$\times \frac{N!}{N_s!(N - N_s)!}$$

$$\cdot e^{-\beta(E_s+E_b)}\delta\left(K_s + K_b + \frac{1}{2}Mc^2\right)$$

$$= \Sigma_{N_s=0}^{N} \int \frac{d^{4N_s}p \, d^{4N_s}q}{N_s! h^{4N_s}} e^{-\beta E_s} Q_{N-N_s}\left(V^{(4)}\right.$$

$$\left. - V_s^{(4)}, T, M + \frac{2K_s}{c^2}\right), \tag{8.58}$$

where we have written the integral measures in a compact form. Let us choose a temperature T so that the principal contributions to the phase space integral come from $E_s \ll E$ and therefore $N_s \ll N$. Then, $K_s \ll K \cong -\frac{1}{2}Mc^2$, and we may approximate

$$Q_{N-N_s}\left((V^{(4)} - V_s^{(4)}), T, M + \frac{2K_s}{c^2}\right)$$

$$\cong \exp\left\{\beta V_s^{(4)} \frac{\partial A}{\partial V^{(4)\prime}}|_{V^{(4)\prime}=V^{(4)}}\right.$$

$$\left. - \beta \frac{2K_s}{c^2}\frac{\partial A}{\partial M'}|_{M'=M} + \beta N_s \frac{\partial A}{\partial N'}|_{N'=N}\right\} \tag{8.59}$$

if the Helmholtz free energy is sufficiently slowly varying at the endpoint values. The first term corresponds to the change in the free energy due to a change in spacetime volume due to the exchange of particles between the subsystem and the bath. We shall assume that the class of subsystems that we are selecting are typical in the sense that they have the same *a priori* time interval associated with them as the bath; therefore the relevant derivative of A holds the time interval fixed, and its associated Δt cancels out. Hence,

$$\frac{\partial A}{\partial V^{(4)}}\Big|_{V^{(4)\prime}=V^{(4)}} = V_s \frac{\partial A}{\partial V'}\Big|_{V'=V} = -V_s P, \qquad (8.60)$$

where V is the spatial volume of the canonical ensemble, and V_s is the spatial volume of the subsystem; the pressure is defined in terms of its usual relation to the Helmholtz free energy.

The second term is an essentially relativistic effect, associated with the mass degree of freedom. We define a corresponding chemical potential

$$\frac{\partial A}{\partial M'}\Big|_{M'=M} = \frac{1}{2}\mu_K c^2, \qquad (8.61)$$

and refer to μ_K as the *mass chemical potential* for the subsystem.

The last term in (8.59) can be immediately identified with the usual chemical potential for the number of particles in the subsystem, so that

$$\frac{\partial A}{\partial N'}\Big|_{N'=N} = \mu. \qquad (8.62)$$

Taking the sum in (8.58) to infinity, we can now define the grand canonical ensemble partition function, providing P, V, T relations as

$$\mathcal{Q}(V^{(4)}, \zeta, z, T) = \Sigma_N z^N \hat{Q}_N(V^{(4)}, \zeta, T) \equiv e^{\beta P V}, \qquad (8.63)$$

where $z = e^{\mu\beta}, \zeta = e^{-\mu_K \beta}$, and

$$\hat{Q}_N(V^{(4)}, \zeta, T) = \int \frac{d^{4N}p\, d^{4N}q}{N! h^{4N}} e^{-\beta E} \zeta^K, \qquad (8.64)$$

and we have dropped the subscript s for the subsystem.

As in the nonrelativistic theory, it follows immediately that

$$\langle N \rangle = z\frac{\partial}{\partial z}\ln \mathcal{Q} = k_B T\frac{\partial}{\partial \mu}\ln \mathcal{Q} \tag{8.65}$$

and that

$$\frac{PV}{k_B T} = \ln \mathcal{Q}(V^{(4)}, \zeta, z, T). \tag{8.66}$$

The average value of K is given by

$$\langle K \rangle = \zeta\frac{\partial}{\partial \zeta}\ln \mathcal{Q} = -k_B T\frac{\partial}{\partial \mu_K}\ln \mathcal{Q}, \tag{8.67}$$

so that it follows from (8.65) and (8.67) that

$$\frac{\partial\langle N \rangle}{\partial \mu_K} = -\frac{\partial\langle K \rangle}{\partial \mu}, \tag{8.68}$$

associating the number dependence on mass and the mass dependence on number in a symmetrical way.

We define the *Helmholtz free energy* for the grand canonical ensemble as

$$A = \langle N \rangle k_B T \ln z + \langle K \rangle k_B T \ln \zeta - k_B T \ln \mathcal{Q}, \tag{8.69}$$

from which it follows that

$$\mathcal{Q} = z^{\langle N \rangle}\zeta^{\langle K \rangle}e^{-\beta A} \tag{8.70}$$

$$1 = \Sigma_N z^{N-\langle N \rangle}e^{\beta A}\int \frac{d^{4n}pd^{4N}d}{N!h^{4n}}e^{-\beta E}\zeta^{K-\langle K \rangle} \tag{8.71}$$

Taking the derivative with respect to β, holding $z, \zeta, V^{(4)}$ fixed, we obtain

$$0 = -\frac{\partial\langle N \rangle}{\partial \beta}\ln z - \frac{\partial\langle K \rangle}{\partial \beta}\ln \zeta + \frac{\partial}{\partial \beta}(\beta A) - \langle E \rangle. \tag{8.72}$$

It follows from (8.69) that

$$\frac{\partial}{\partial \beta}(\beta A) = \frac{\partial\langle N \rangle}{\partial \beta}\ln z + \frac{\partial\langle K \rangle}{\partial \beta}\ln \zeta - \frac{\partial}{\partial \beta}\ln \mathcal{Q}, \tag{8.73}$$

so that

$$U \equiv \langle E \rangle = -\frac{\partial}{\partial \beta} \ln Q, \qquad (8.74)$$

and hence, from the definitions of A and U,

$$U = A - \mu \langle N \rangle + \mu_K \langle K \rangle + k_B T \ln Q + k_B T^2 \frac{\partial}{\partial T} \ln Q. \qquad (8.75)$$

from the relation $U = A + TS$, we identify the entropy

$$S = \frac{\partial}{\partial T}(k_B T \ln Q) + \frac{\mu_K \langle K \rangle}{T} - \frac{\mu \langle N \rangle}{T}. \qquad (8.76)$$

With these results, it is easy to verify the Maxwell relations

$$S = -\left(\frac{\partial A}{\partial T}\right)_{V, \langle N \rangle, \langle K \rangle},$$

$$P = -\left(\frac{\partial A}{\partial V}\right)_{T, \langle N \rangle, \langle K \rangle}, \qquad (8.77)$$

and to verify, in terms of our original definitions, that

$$\mu = \left(\frac{\partial A}{\partial N}\right)_{\langle K \rangle, T, V}$$

$$\mu_K = -\left(\frac{\partial A}{\partial \langle K \rangle}\right)_{\langle N \rangle, T, V}. \qquad (8.78)$$

If the grand canonical ensemble peaks strongly at a particular value of N, the canonical ensemble is recovered for that N, and similarly for a particular value $\bar{K}$ of K, which we could call $-\frac{1}{2}\bar{M}c^2$.

Using the asymptotic form for E previously studied, the Galilean limit of the grand canonical ensemble can be obtained for each N using the relation

$$e^{-\beta E} e^{-\beta \mu_K K} \sim e^{-\beta(\varepsilon' + Mc^2)} e^{-\beta \mu_K (H - \varepsilon' - Mc^2/2)}$$

$$= e^{-\beta Mc^2} e^{\beta \mu_K Mc^2/2} e^{\beta \varepsilon'(1 - \mu_K)} e^{-\beta \mu_K H}, \qquad (8.79)$$

from which we see that $\mu_K \to 1$ as $c \to \infty$. Then the ε' dependence cancels, and the first two factors combine to $e^{-\beta Mc^2/2}$.

The exponential damping factor is compensated if

$$\mu \sim \mu_k + \frac{1}{2}\frac{M}{N}c^2, \tag{8.80}$$

where the quantity M/N, the average mass per particle, should not depend on N. With these assumptions, and the approximation $dE_i/c \sim cdm_i$, one finds that

$$\mathcal{Q}(V^{(4)}, \zeta, z, T) \sim \Sigma_0^\infty z'^N \hat{Q}_N(V, T) \tag{8.81}$$

where $z' = e^{\mu_{NR}\beta}$, and

$$\hat{Q}_N(V, T) = \int \frac{d^{3N}p\, d^{3N}q}{N! h^{3N}} e^{-\beta H}, \tag{8.82}$$

and we have taken the limit in $\Delta m, \Delta t$ integrations so that

$$c\Delta mc\Delta t/h = \Delta \eta \Delta t/h = 1. \tag{8.83}$$

Using the general formulas for the dispersions from the grand canonical ensemble, by differentiating with respect to μ_K and μ,

$$\langle (K - \langle K \rangle)^2 \rangle = -k_B T \frac{\partial \langle K \rangle}{\partial \mu_K},$$

$$\langle (N - \langle N \rangle)^2 \rangle = k_B T \frac{\partial \langle N \rangle}{\partial \mu} \tag{8.84}$$

for the free particle gas, one finds that [Horwitz (1981)]

$$\frac{\Delta K_0}{\langle K_0 \rangle} = \frac{1}{\sqrt{\langle N \rangle}} + O\left(\frac{1}{c^4}\right), \tag{8.85}$$

just as for the fluctuations in N, thus verifying the close relationship between K and the number of particles. In a similar way, one can verify that [Horwitz (1981)]

$$\frac{\langle E \rangle - M_0 c^2 \langle N \rangle}{\langle N \rangle} = \frac{3}{2}k_B T + O\left(\frac{1}{c^2}\right) \tag{8.86}$$

and the the Dulong-Petit law for the specific heat becomes

$$\frac{\partial}{\partial T}\frac{\langle E \rangle - M_0 c^2 \langle N \rangle}{\langle N \rangle} = \frac{3}{2}k_B + O\left(\frac{1}{c^2}\right). \tag{8.87}$$

Therefore, the relativistic grand canonical ensemble that we have developed here contains the standard results of the nonrelativistic theory, and can serve as a useful generalization with applications from classical microscopic systems to astronomical calculations [Hakim (2011)].

8.5 Relativistic Quantum Statistical Mechanics

In this section we show how the results of the previous sections on classical statistical ensembles can be extended to the quantum case. In the quantum theory, the density operator for a quantum state composed of a maximal mixture of energies, which corresponds to a microcanonical ensemble, is represented by

$$\rho = \Sigma_{k,E\in\Delta}\psi_{k,E}\psi^{*}_{k,E}, \tag{8.88}$$

where $\psi_{k,E}$ are eigenfunctions (or generalized eigenfunctions in the case of continuous spectrum) of the total K operator, and E is the value of the total energy operator (well-defined by translation invariance of the whole system), as discussed for the classical microcanonical ensemble. The masses m_i in this representation may be specified to lie in small intervals μ_i as for the classical case discussed above. Then, the total number of states with $k, E \in \Delta$ and $m_i \in \mu_i$ is

$$\Gamma(k,E) = Tr\rho = \Sigma_{k,E\in\Delta}\|\psi_{k,E}\|^{2}, \tag{8.89}$$

and the entropy is defined as

$$S(k,E) = k_B \ln \Gamma(k,E). \tag{8.90}$$

The canonical ensemble is defined as for the classical case discussed above, and is given by

$$\rho = e^{-\beta E}\delta_{K,-Mc^2/2}, \tag{8.91}$$

where $\beta = 1/k_B T$, and the partition function is

$$Q_N(V^{(4)}, T, M) = Tr(e^{\beta E}\delta_{K,-Mc^2/2}), \tag{8.92}$$

so that the expectation of any operator $\mathcal{O}$ in this ensemble is given by

$$\langle \mathcal{O} \rangle_N = \frac{Tr(\mathcal{O}e^{\beta E}\delta_{K,-Mc^2/2})}{Q_N}. \tag{8.93}$$

To obtain the grand canonical ensemble, we select a subsystem s of the system described by a canonical ensemble of N particles, and write the partition function, as done for the classical case, as in (8.38), as

$$Q_N(V^{(4)}, T, M) = \Sigma_{N_s+N_b=N} Tr_s Tr_b(e^{-\beta E_s} e^{-\beta E_b} \delta_{K_s+K_b,-Mc^2/2})$$
$$= \Sigma_{N_s=0}^N Tr_s(V^{(4)} - V_s^{(4)}, T, M + 2K/c^2). \tag{8.94}$$

Following the procedure of the classical case, we obtain (8.70), *i.e.*,

$$\mathcal{Q}(V^{(4)}, \zeta, z, T) = \Sigma_N z^N \hat{Q}_N(V^{(4)}, \zeta, T) \equiv e^{\beta PV}, \tag{8.95}$$

where now

$$\hat{Q}_N(V^{(4)}, \zeta, T) = Tr_N(e^{-\beta E}\zeta^K), \tag{8.96}$$

and, as we shall see, $\zeta = e^{-\mu_K \beta}$.

For the ideal free quantum gas in a spacetme box (defined as above) of dimension

$$-L/2 \le x, y, z \le L/2, -\Delta t/2 \le t \le \Delta t/2,$$

the microcanonical distribution is characterized by the spectrum

$$2MK = \hbar^2(k_1^2 + k_2^2 + k_3^2 - k_0^2) \tag{8.97}$$

where

$$k_0 = \frac{2\pi}{\Delta t}\nu_0, \quad k_j = \frac{2\pi\nu_j}{L} \quad j = 1, 2, 3, \tag{8.98}$$

and $\nu_0, \nu_j = 0, \pm 1, \pm 2 \dots$. Then, $\mathbf{p} = (2\pi\hbar/L)\nu$ and $\varepsilon = (2\pi\hbar/\Delta t)\nu_0$. The integral measure is given by

$$d^3p\, d\varepsilon \sim \frac{(2\pi\hbar)^4}{V^{(4)}}, \quad V^{(4)} = L^3 \Delta t. \tag{8.99}$$

We now compute the Bose-Einstein, Fermi-Dirac and Boltzmann distributions in terms of the discrete sums characteristic of kinetic theory. Let

$i = $ cell around $\mathbf{p}, \varepsilon, m \in \mu,$

$g_i = $ number of mass and momentum states in each cell   (8.100)

$n_i = \Sigma_{\mathbf{p},\varepsilon} n_{\mathbf{p},\varepsilon},$

where $n_{\mathbf{p},\varepsilon}$ is the number of particles with energy momentum $\mathbf{p}, \varepsilon$. Let $W(\{n_i\})$ be the number of states associated with the distribution $\{n_i\}$. Then, the total number of states in phase space is

$$\Gamma(E, K_0) = \Sigma_{\{n_i\}} W(\{n_i\}) \qquad (8.101)$$

with the constraints

$$E = \Sigma_i \varepsilon_i n_i \quad K_0 = \Sigma_i K_i n_i \quad N = \Sigma_i n_i, \qquad (8.102)$$

where K_i is the average value of $K(\mathbf{p}, \varepsilon)$ in the ith cell. Taking into account the constraints (8.102) we wish to find $\{n_i\}$ such that

$$\delta\{\ln W(\{n_i\}) - \alpha\Sigma_i n_i - \beta\varepsilon_i n_i + \gamma\Sigma_i K_i n_i\} = 0 \qquad (8.103)$$

where α, β, γ are Lagrange parameters implementing the constraints. Permitting up to g_i states in each cell for Fermi–Dirac statistics, and all integer values for Bose–Einstein statistics, one finds the distributions

$$W(\{n_i\}) = \Pi_i \frac{(n_i + g_i - 1)!}{N_i!(g_i - 1)!} \quad (\text{Bose} - \text{Einstein})$$

$$= \Pi_i \frac{g_i!}{n_i!(g_i - n_i)!} \quad (\text{Fermi} - \text{Dirac})$$

$$= \Pi_i \frac{g_i!}{n_i!} \quad (\text{Boltzmann}) \qquad (8.104)$$

and obtains the average occupation number distributions (the sign of K_i is important in establishing the sign of the second variation)

$$\bar{n}_i = \frac{g_i}{z^{-1}\zeta^{-K_i}e^{\beta\varepsilon_i} - 1} \quad (\text{Bose} - \text{Einstein})$$

$$= \frac{g_i}{z^{-1}\zeta^{-K_i}e^{\beta\varepsilon_i} + 1} \quad (\text{Fermi} - \text{Dirac})$$

$$= g_i z \zeta^{K_i} e^{-\beta\varepsilon_i} \quad (\text{Boltzmann}), \qquad (8.105)$$

where $z = e^\alpha$ and $\zeta = e^\gamma$. Using the maximal distributions in (8.101), the entropy is given by

$$S = k_B \ln W(\{\bar{n}_i\}). \qquad (8.106)$$

Pinching down the size of the cells to obtain continuum distributions, we can write (taking $g_i = 1$)

$$\bar{n}_{\mathbf{p},\varepsilon} = \frac{1}{z^{-1}\zeta^{-K(\mathbf{p},\varepsilon)}e^{\beta\varepsilon} - 1} \qquad \text{(Bose} - \text{Einstein)}$$

$$= \bar{n}_{\mathbf{p},\varepsilon}$$

$$= \frac{1}{z^{-1}\zeta^{-K(\mathbf{p},\varepsilon)}e^{\beta\varepsilon} + 1} \qquad \text{(Fermi} - \text{Dirac)}$$

$$= z\zeta^{K(\mathbf{p},\varepsilon)}e^{-\beta\varepsilon} \qquad \text{(Boltzmann)}. \qquad (8.107)$$

The parameters z, β, ζ are to be determined from

$$\Sigma_{\mathbf{p},\varepsilon}\varepsilon\bar{n}_{\mathbf{p},\varepsilon} = E,$$

$$\Sigma_{\mathbf{p},\varepsilon}\bar{n}_{\mathbf{p},\varepsilon} = N, \qquad (8.108)$$

$$\Sigma_{\mathbf{p},\varepsilon}K(\mathbf{p},\varepsilon)\bar{n}_{\mathbf{p},\varepsilon} = K_0,$$

where the sums are to be taken over a narrow range of masses Δm. Comparing, in the Boltzmann case, with the classical grand canonical distributions, we identify $\beta = \frac{1}{k_B T}, z = e^{\mu\beta}, \zeta = e^{-\mu_K \beta}$ (we show below that these results can be derived directly from the quantum grand canonical ensemble).

Note that in the Fermi–Dirac distribution we have counted as distinct states the several values of ε for each $\mathbf{p}$ which lie within the admissible width Δm. Although the distributions we have obtained are formally very similar (except for the factor ζ^K), the usual notion of the Fermi-Dirac statistics treats all of these states as the same; the role of the mass potential μ_K is to control this mass distribution within the small interval Δm, and thus it is expected that the results of any expectation value remain, in general, closely the same as in the mass shell theories.

Using Stirling's approximation for the factorials, one finds that for the Boltzmann gas,

$$S/k_B = \beta E - K \ln \zeta - N \ln z. \qquad (8.109)$$

We now turn to a study of the ideal gas from the point of view of the grand canonical ensemble.

For Boltzmann statistics, (8.64) can be written as

$$\hat{Q}_N(V^{(4)}, \zeta, T) = \Sigma_{n_{\mathbf{p},\varepsilon}} \frac{1}{N!} \left(\frac{N!}{\Pi_{\mathbf{p},\varepsilon} n_{\mathbf{p},\varepsilon}} \right) e^{-\beta E(\mathbf{p},\varepsilon)} \zeta^{K(\mathbf{p},\varepsilon)}, \qquad (8.110)$$

where

$$E(\{n_{\mathbf{p},\varepsilon}\}) = \Sigma_{\mathbf{p},\varepsilon} \varepsilon n_{\mathbf{p},\varepsilon},$$
$$K(\{n_{\mathbf{p},\varepsilon}\}) = \Sigma_{\mathbf{p},\varepsilon} K(\mathbf{p}, \varepsilon) \qquad (8.111)$$

and

$$N = \Sigma_{\mathbf{p},\varepsilon} n_{\mathbf{p},\varepsilon} \qquad (10.129)$$

as a constraint.

With the constraint (8.112), the sum in (8.110) becomes

$$\hat{Q}_N(V^{(4)}, \zeta, T) = \frac{1}{N!} \left(\Sigma_{\mathbf{p},\varepsilon} e^{-\beta \varepsilon} \zeta^{K(\mathbf{p},\varepsilon)} \right)^N, \qquad (8.113)$$

and therefore

$$\mathcal{Q}(V^{(4)}, \zeta, z, T) = \exp \left\{ z \Sigma_{\mathbf{p},\varepsilon} e^{-\beta \varepsilon} \zeta^{K(\mathbf{p},\varepsilon)} \right\}. \qquad (8.114)$$

The equation of state can then be obtained explicitly by noting that, as in the classical case,

$$\langle N \rangle = z \frac{\partial}{\partial z} \ln \mathcal{Q} = \ln \mathcal{Q} = \frac{PV}{k_B T}. \qquad (8.115)$$

For Bose-Einstein and Fermi-Dirac statistics the distribution functions become [Horwitz (2015)]

$$\mathcal{Q}(V^{(4)}, \zeta, z, T) = \Pi_{\mathbf{p},\varepsilon} \frac{1}{1 - z e^{-\beta \varepsilon} \zeta^{K(\mathbf{p},\varepsilon)}} \quad (\text{Bose} - \text{Einstein})$$

$$= \Pi_{\mathbf{p},\varepsilon} (1 + z e^{-\beta \varepsilon} \zeta^{K(\mathbf{p},\varepsilon)}) \quad (\text{Fermi} - \text{Dirac}).$$
$$(8.116)$$

The equations of state for the relativistic free quantum gas are then

$$\frac{PV}{k_B T} = \ln \mathcal{Q} = -\Sigma_{\mathbf{p},\varepsilon} \ln(1 - ze^{-\beta\varepsilon}\zeta^{K(\mathbf{p},\varepsilon)}) \quad (\text{Bose} - \text{Einstein})$$

$$= \Sigma_{\mathbf{p},\varepsilon} \ln(1 + ze^{-\beta\varepsilon}\zeta^{K(\mathbf{p},\varepsilon)}) \quad (\text{Fermi} - \text{Dirac}). \tag{8.117}$$

The total number of particles is

$$N = z\frac{\partial}{\partial z}\ln \mathcal{Q} = \Sigma_{\mathbf{p},\varepsilon}\frac{ze^{-\beta\varepsilon}\zeta^{K(\mathbf{p},\varepsilon)}}{1 - ze^{-\beta\varepsilon}\zeta^{K(\mathbf{p},\varepsilon)}} \quad (\text{Bose} - \text{Einstein})$$

$$= \Sigma_{\mathbf{p},\varepsilon}\frac{ze^{-\beta\varepsilon}\zeta^{K(\mathbf{p},\varepsilon)}}{1 + ze^{-\beta\varepsilon}\zeta^{K(\mathbf{p},\varepsilon)}} \quad (\text{Fermi} - \text{Dirac}) \tag{8.118}$$

Similarly, by differentiating with respect to $\beta\varepsilon - \ln K(\mathbf{p}, \varepsilon)$, we find that the average occupation numbers are given by

$$\langle n_{\mathbf{p},\varepsilon}\rangle = \frac{ze^{-\beta\varepsilon}\zeta^{K(\mathbf{p},\varepsilon)}}{1 \mp ze^{-\beta\varepsilon}\zeta^{K(\mathbf{p},\varepsilon)}}. \tag{8.119}$$

(8.118) then correspond to

$$N = \Sigma_{\mathbf{p},\varepsilon}\langle n_{\mathbf{p},\varepsilon}\rangle. \tag{8.120}$$

8.6 Relativistic High Temperature Boson Phase Transition

Haber and Weldon [Haber (1982)] showed that in the usual (mass shell) form of relativistic quantum mechanics, taking into account both the particle and antiparticle distribution functions, a system of bosons can undergo a high temperature phase transition. The introduction of antiparticles in the theory, by application of the arguments of Haber and Weldon, imply the addition of another term in the total number expectation with a negative sign, carrying an oppposite sign for the energy chemical potential, *i.e.*, formula (8.118) (for the boson case) is written as [Burakovsky (1996)](dividing numerator and denominator by the numerator factor)[2]

$$N = V^{(4)}\Sigma_{k\mu}\left[\frac{1}{e^{(E-\mu-\mu_K m^2/2M/T)}-1}\right.$$
$$\left.-\frac{1}{e^{(E+\mu-\mu_K m^2/2M/T)}-1}\right].$$ (8.121)

As assumed by Haber and Weldon, the *total* particle number remains unchanged in the equilibrium state, but the presence of antiparticles implies annihilation and creation processes. Thus, in counting the total number of particles, the antiparticle distribution must carry a negative sign, consistent with the interpretation of Stueckelberg as given in the earlier chapters of this book. On the other hand, both the terms in the sum in (8.121) must separately be positive, implying the inequalities

$$m - \mu - \mu_K\frac{m^2}{2M} \geq 0,$$
$$m + \mu - \mu_K\frac{m^2}{2M} \geq 0,$$ (8.122)

resulting in the inequalities representing the non-negativeness of the discriminants in the mass quadratic formulas,

$$-\frac{M}{2\mu_K} \leq \mu \leq \frac{M}{2\mu_K}.$$ (8.123)

The bounds of the intersection of the regions satisfying the inequalities (8.121) are given by

$$\frac{M}{\mu_K}\left(1-\sqrt{1-\frac{2|\mu|\mu_K}{M}}\right) \leq m \leq \frac{M}{\mu_K}\left(1+\sqrt{1-\frac{2|\mu|\mu_K}{M}}\right),$$ (8.124)

which for small $\frac{|\mu|\mu_K}{M}$ reduces, as in the no-antiparticle case, to

$$|\mu| \leq m \leq \frac{2M}{\mu_K}.$$ (8.125)

---

[2]Since the sign of the energy of the antiparticle is opposite to that of the particle, the chemical potential μ must change sign for the antiparticle. However the mass squared of both particle and antiparticle are positive, and therefore the sign of μ_K does not change.

Replacing the summation in (8.121) by integration, one obtains the formula for the number density

$$n = \frac{1}{4\pi^3} \int_{m_1}^{m_2} m^3 dm \int_{-\infty}^{\infty} \sinh^2 \beta d\beta$$

$$\times \left[\frac{1}{e^{(m\cosh\beta - \mu - \mu_K m^2/2M)/T} - 1} \right.$$

$$\left. - \frac{1}{e^{(m\cosh\beta + \mu - \mu_K m^2/2M)/T} - 1} \right] \tag{8.126}$$

where m_1 and m_2 are defined by the bounds in (8.24). Integrating out the β variable, one finds for high temperature $\mu/T \ll 1$,

$$n \cong \frac{1}{\pi^3} \left(\frac{M}{\mu_K} \right)^2 \mu T \sqrt{1 - \frac{2|\mu|\mu_K}{M}}. \tag{8.127}$$

For T above a critical value, the range of admissible masses become pinched down to zero, corresponding to a phase transition, where the dispersion

$$\delta m = \sqrt{\langle m^2 \rangle - \langle m \rangle^2}$$

vanishes as $\sqrt{T - T_c}$, a second order transition, corresponding to a ground state with $p_\mu p^\mu = -(M/\mu_k)^2$. States with temperature $T > T_c$ correspond to off-shell excitations of such a ground state.

The phase transition that we have described selects a definite mass for the particles, but this result is statistical. Although the mean fluctuations vanish, there is nevertheless sufficient freedom in the phase space for each particle to fulfil the off-shell requirements for the formulation of the Stueckelberg theory.

8.7 Black Body Radiation

As we have discussed in previous chapters, the Stueckelberg-Schrödinger equation implies that the electromagnetic gauge fields are five dimensional, including an a_5 field which compensates for the τ derivative of the evolving wave function. The usual argument for two polarization states of the four dimensional Maxwell field is that, of the four degrees of freedom, there is a gauge condition and the constraint of the Guass law, leaving two polarization states.

The factor of two on the Bose-Einstein distribution is essential for the computation of the specific heat of a black body, but the argument of the existence of two constraints leaves the possibility of three polarization states. In the following, we show that the observable radiation field of a black body indeed carries just two polarization states [Horwitz (2015)].

The canonical quantization of the 5D radiation field was carried out by Shnerb and Horwitz [Shnerb (1993)] following the basic ideas of Teitelboim and Henneaux [Henneaux (1992)] and Haller [Haller (1972)] using algebraic methods. Taking for this discussion, as in [Shnerb (1993)], the signature of the 5D manifold to be $[\sigma, +, -, -, -]$, we write the action for the interacting fields (for both quantized gauge fields and quantized wave functions ψ) as

$$
\begin{aligned}
S = \int_{-\infty}^{\infty} d^5x \bigg\{ &-\frac{\lambda}{4} f^{\alpha\beta} f_{\alpha\beta} - G(x)[\partial_\alpha a^\alpha(x)] \\
&+ \frac{1}{2\lambda} G^2(x) \left(i\psi^\dagger(x) \frac{\partial \psi(x)}{\partial \tau} \right) - \frac{1}{2M} \psi^\dagger(x) \frac{\partial \psi(x)}{\partial \tau} \\
&- \frac{1}{2M} \psi^\dagger [\partial^\mu - ie'a^\mu(x)][\partial_\mu - ie'a_\mu(x)]\psi(x) \\
&+ e'\psi^\dagger(x) a_\tau(x) \psi(x) \bigg\},
\end{aligned}
\tag{8.128}
$$

where λ is a quantity with dimensions of length. As discussed in Chapter 7, e' is the coupling constant of the covariant theory, which also has dimension of length, and G plays the role of an auxiliary field [Haller (1972)] (somewhat analogous to the Faddeev-Popov ghosts [Faddeev (1967)] of the path integral approach). The canonically conjugate momenta are given by

$$
\pi^\mu = \frac{\delta \mathcal{L}}{\delta(\partial_\tau a_\mu)} = -\lambda f^{\tau\mu},
$$

$$
\pi^\tau = \frac{\delta \mathcal{L}}{\delta(\partial_\tau a_\tau)} = -\sigma G,
\tag{8.129}
$$

$$
\pi_\psi = \frac{\delta \mathcal{L}}{\delta(\partial_\tau \psi)} = i\psi^\dagger.
$$

We now impose equal time commutation relations

$$[\pi^\alpha(x), a_\beta(y)] = -i\delta_\beta^\alpha \delta(x-y) \qquad (8.130)$$

and (we are assuming ψ a boson field for our present purposes)

$$[i\psi^\dagger(x), \psi(y)] = -i\delta(x-y). \qquad (8.131)$$

The Hamiltonian (generating unitary evolution in ψ and a^α then takes the form

$$K = \sigma \int d^4x [\pi^\mu(\partial_\tau a_\mu) + \pi^\tau(\partial_\tau a_\tau) + i\psi^\dagger \partial_\tau \psi - \mathcal{L}]$$

$$= K_\gamma + K_m + K_{\gamma m}, \qquad (8.132)$$

where

$$K_\gamma = \int d^4x \left\{ -\frac{1}{2\lambda}\pi^\mu \pi_\mu - \frac{\lambda\sigma}{4}f^{\mu\nu}f_{\mu\nu} \right.$$

$$\left. + \pi^\mu(\partial_\mu a^\tau) - \pi^\tau(\partial_\mu a^\mu) - \frac{1}{2\lambda}\pi^\tau \pi_\tau \right\} \qquad (8.133)$$

and

$$K_m = \frac{\sigma}{2M} \int d^4x \psi^\dagger \partial_\mu \partial^\mu \psi,$$

$$K_{\tau m} = d^4x \left\{ -e'\psi^\dagger a_\tau \psi - \frac{ie'}{2M}\psi^\dagger[a^\mu\partial_\mu + (\partial_\mu a^\mu)] \right.$$

$$\left. - \frac{e'^2}{2M}\psi^\dagger \psi a^\mu a_\mu \right\}. \qquad (8.134)$$

The stability condition on the states for the restriction to the Gauss law

$$\langle \partial_\mu \pi^\mu + j^\tau \rangle = 0 \qquad (8.135)$$

implies that $\langle \pi^\tau \rangle = 0$; one can then eliminate the longitudinal part of the field a^μ.

In case the four vector k^μ in the Fourier decomposition of the a^μ field is timelike, for which the $O(4,1)$ theory is stable, one can eliminate, by a unitary transformation (as in the Maxwell case), the time component of a^μ. There remains, except for the Coulomb term, three spacelike polarization components a^i, and the Hilbert space has positive norm.

We have argued in Chapter 4 that vector bosons must lie in a representation of angular momentum with spin 1; as discussed in Jauch and Rohrlich [Jauch (1955) p. 41], these components, with canonical commutation relations, provide a representation, in any choice of gauge, that meets this requirement. For the asymptotic photons of the black body radiation, the components for k^μ spacelike, for which the stable solutions are representations of $O(2,1)$ do not meet this requirement. Furthermore, in the case that k^μ is lightlike, the elimination of longitudinal modes corresponds exactly to the removal of both a_0 and $a_\parallel$, leaving just two polarization states. This limiting case is realized for the asymptotic photons of the black body when $\tau \to \infty$, leaving, by application of the Riemann-Lebesgue lemma, the "massless" zero mode. We make this argument explicit in the following.

The analog of the radiation gauge (*e.g.* Bjorken (1964))for the $5D$ fields would correspond to setting the a_5 field equal to zero; this corresponds to subtracting the 5-gradient of the indefinite integral of the a_5 field from the a_α fields, *i.e.*, for

$$a^{5'} = a^5 + \partial^5 \Lambda, \tag{8.136}$$

we can take

$$\Lambda(x,\tau) = -\int^\tau a^5(x,\tau') + \tilde{\Lambda}(x). \tag{8.137}$$

Then, since the second term is independent of τ, $a^{5'} = 0$. Furthermore, since

$$a^{\mu'} = a^\mu + \partial^\mu \Lambda(x), \tag{8.138}$$

it follows that

$$a^{0'} = a^0(x,\tau) - \partial^0 \int^\tau a^5(x,\tau')d\tau' + \partial^0 \tilde{\Lambda}(x). \tag{8.139}$$

Under the assumption that the asymptotic fields are independent of τ, assuming as well convergence of the indefinite integral in (8.139) for large τ, we can make $a^{0'} = 0$ asymptotically with the choice

$$\tilde{\Lambda}(x) = -\int^t a^0(\mathbf{x},t',\tau))dt' + \int^t \int^\tau a^5(\mathbf{x},t',\tau')d\tau'dt'. \tag{8.140}$$

The remaining term of the generalized Lorentz gauge $\partial_\alpha a^\alpha = 0$ is just the condition $\nabla \cdot \mathbf{a} = 0$, exhibiting the required rotational invariance on the orbit of the induced representation for the a^μ field. The longitutinal component along the $\mathbf{k}$ vector must therefore vanish, and we are left with two effective polarization states.

Therefore, with the Gauss law and the additional gauge condition on the $5D$ fields, there are three constraints on the $5D$ fields, leaving two degrees of freedom.

The remaining degrees of freedom correspond, in the induced representation, to two polarization states that are directly interpretable as angular momentum states of the photon in $SU(2)$ on the orbit.

The boson distribution function obtained above with the remaining two degrees of freedom, then gives the usual result for the specific heat for black body radiation [Horwitz (2015)].

We remark that the relativistic Gibbs ensembles worked out above assumed, for simplicity, that there were no antiparticles (the Boltzmann counting construction did not make this assumption). The existence of the a_5 field makes possible, as we have seen in Chapter 7, the (classical) particle-antiparticle transition on particle world lines. The analog of the radiation gauge requirement that we have imposed above as a second gauge condition, resulting in residually two degrees of freedom for the radiation field, would not admit this mechanism in the detectors. The presence of pair production (expected to be very small [Schwinger (1951)] in the detector would therefore suggest that there may be this additional degree of freedom in the boson gas, with a concomitantly small correction in the black body radiation formula.

8.8 Manifestly Covariant Relativistic Boltzmann Equation

In this section, we shall derive a covariant Boltzmann equation with collision terms obtained from the binary scattering of events as described by relativistic scattering theory. We give here the basic ideas, and refer the reader to the work of Horwitz, Shashoua and Schieve [Horwitz (1981)] for details.

We study the case of N identical particles, and use, for convenience, the formalism of second quantization. The field which annihilates an event at the point $q = (\mathbf{q}, t)$ is related to the operator which annihilates an event of energy momentum $p = (\mathbf{p}, E/c)$ by the Fourier transform ($\hbar = 1$)

$$\psi(q) = \frac{1}{(2\pi)^2} \int d^4p\, \psi(p) e^{ip \cdot q}. \tag{8.141}$$

An arbitrary operator A on the Hilbert space of events can be represented as

$$A = \Sigma_{s=1}^{N} \frac{1}{s!} \int d^4q_1 \ldots d^4q_s \psi^\dagger(q_1) \cdots \psi^\dagger(q_s) \hat{A}_s \psi(q_1) \cdots \psi(q_s),$$

$$\tag{8.142}$$

where $\hat{A}$ are operators acting on the space associated with every s-event subspace of the N event system. The expectation value of such an operator can be expressed in terms of the trace with the density matrix ρ as

$$\langle A \rangle = Tr(\rho A). \tag{8.143}$$

The Weyl correspondence [Weyl (1931)] applies, as in the nonrelativistic theory, to every s-event operator represented as [Balescu (1975)]

$$\hat{A}_s = \int d^4k_1 d^4j_1 \ldots d^4k_s d^4j_s A_s(k_1 j_1 \cdots k_s j_s)$$

$$\times \exp\left\{ i\Sigma_{n=1}^{s} k_n \cdot \hat{q}_n + j_n \cdot \hat{p}_n \right\}, \tag{8.144}$$

where the operators $\hat{q}_n, \hat{p}_n$ satisfy the canonical commutation relations

$$[\hat{q}_n^\mu, \hat{p}_{n'}^\nu] = ig^{\mu\nu} \delta_{n,n'} \tag{8.145}$$

There is a corresponding function $A_s(q_1, p_1, \ldots q_s p_s)$ of the classical variables containing the same coefficients $A_s(k_1 j_1 \cdots k_s j_s)$ which is its classical limit. Consider, in particular, the case $s = 1$. Then, the quantity $\langle A_1 \rangle$ is given by

$$\langle A_1 \rangle = \int d^4q \int d^4k d^4j A_1(k,j) Tr(\rho \psi^\dagger(q) e^{i(k \cdot \hat{q} + j \cdot \hat{p})} \psi(q)). \quad (8.146)$$

The exponential can be factorized to

$$\exp(ik \cdot q + j\partial) = \exp(ik \cdot q) \exp(j \cdot \partial) \exp(ik \cdot j/2). \quad (8.147)$$

Then (8.139) becomes

$$\langle A_1 \rangle = \int d^4q d^4p A_1(q,p) f_1^W(q,p), \quad (8.148)$$

where $A_1(q,p)$ is the classical function corresponding to the operator $\hat{A}_1$ through the Weyl correspondence, and we have defined the *one particle relativistic Wigner function*

$$f_1^W(q,p) = \frac{1}{(2\pi)^4} \int d^4j e^{-ij \cdot p} Tr\left(\rho \psi^\dagger(q - \frac{j}{2}) \psi\left(q + \frac{j}{2}\right)\right)$$

$$= \frac{1}{(2\pi)^4} \int d^4k e^{ik \cdot q} Tr\left(\rho \psi^\dagger(p - \frac{k}{2}) \psi\left(p + \frac{k}{2}\right)\right). \quad (8.149)$$

As for the nonrelativistic analog of this procedure, $f_1^W(q,p)$ is not necessarily positive, and cannot be interpreted as a pointwise probability density. It has the advantage, as we shall see, that the equations of motion are very analogous to the classical equations in phase space, and the results are immediately applicable to classical transport theory. Furthermore, note that

$$\int d^4q f_1^W(q,p) = Tr(\rho \psi^\dagger(p) \psi(p)) \geq 0, \quad (8.150)$$

and that

$$\int d^4q d^4p f_1^W(q,p) = \int d^4q Tr(\rho \psi^\dagger(q) \psi(q)). \quad (8.151)$$

Since

$$\int d^4q \psi^\dagger(q) \psi(q) = \int d^4q \psi^\dagger(p) \psi(p) = N, \quad (8.152)$$

the number operator for the total absolutely conserved number of the set of events, is a superselection rule [Piron (1976)] for this system, and therefore just a simple classical number,

$$\int d^4q d^4p f_1^W(q,p) = NTr\rho = N, \qquad (8.153)$$

i.e. a "normalization" for the Wigner function.

We now consider the τ evolution of the one particle distribution function. To do this in a convenient way, we study the Fourier transform

$$f_1^W(k,p) = \int d^4q e^{ik\cdot q} f_1^W(q,p) = Tr\left(\rho\psi^\dagger\left(p-\frac{k}{2}\right)\psi\left(p+\frac{k}{2}\right)\right). \qquad (8.154)$$

Using the cyclic properties of operators under a trace with the density matrix, it then follows from the Stueckelberg-Schrödinger evolution that

$$\partial_\tau f_1^W(k,p) = -iTr\left[\left(\rho\psi^\dagger\left(p-\frac{k}{2}\right)\psi\left(p+\frac{k}{2}\right)\right), K\right]. \qquad (8.155)$$

We assume that K has the form

$$K = K_0 + V \qquad (8.156)$$

where

$$K_0 = -\int d^4q \psi^\dagger(q)\frac{\partial^\mu\partial_\mu}{2M}\psi(q), \qquad (8.157)$$

and

$$V = \frac{1}{2}\int d^4q' d^4q'' \psi^\dagger(q')\psi^\dagger(q'')V(q'-q'')\psi(q'')\psi(q') \qquad (8.158)$$

is the two body operator (Poincaré invariant) corresponding to a two-event interaction potential. Carrying out the commutator with this model, one finds that the time dependence of the one particle Wigner function depends on the two particle Wigner function, defined by

$$f_2^W(k_1 p_1, k_2 p_2) = \int d^4 q_1 d^4 q_2 e^{-ik_1 \cdot q_1 - ik_2 \cdot q_2} f_2^W(q_1 p_1, q_2 p_2)$$

$$= Tr(\rho \psi^\dagger(p_1 - k_1/2) \psi^\dagger(p_2 - k_2/2)$$

$$\times \psi(p_1 + k_1/2) \psi(p_2 + k_2/2)), \qquad (8.159)$$

according to

$$\partial_\tau f_1^W(k_1, p_1) = L_0 f_1^W(k_1, p_1)$$

$$+ \int d^4 p_2 d^4 k_2 \delta^4(k_2) L_{12} f_2^W(k_1 p_1, k_2 p_2), \qquad (8.160)$$

where L_0 and L_{12} are differential operators induced by the commutator with K_0. This procedure may be applied again to every f_s^W for $s = 1, 2 \ldots N$, and results in a set of equations of precisely the same form as the well-known BBGKY hierarchy [Balescu (1975)] for the nonrelativistic case. One obtains in this way a relativistically covariant generalization of the BBGKY hierarchy derived from basic dynamical principles.

The higher order relations invoke higher order correlations, and for a dilute gas of events, we may assume that truncation at the level of two body correlations will suffice. Furthermore, the two body correlation terms can be represented to fairly good accuracy, as in the non-relativistic case, by two body scattering amplitudes, consisting of two basic terms, one scattering events into the quasi-equilibrium ensemble, and the other, scattering events out. The basic ingredients needed are derived in Chapter 6 on scattering theory. The scattering, as for the nonrelativistic case, induces changes in the distribution function, *i.e.*, the rate of change of f due to collisions is

$$D_c f(q, p) = D_c^+ f(q, p) - D_c^- f(p, q) \qquad (8.161)$$

where $D_c^- f(p, q) d^4 q d^4 p \delta \tau$ is the number of collisions in the interval $\delta \tau$ in which one of the events is in $d^4 q d^4 p$, and $D_c^+ f(p, q) d^4 q d^4 p \delta \tau$ is the number of collisions in $\delta \tau$ in which one of the final events is in $d^4 q d^4 p$. Denoting by $\dot{P}$ the transition rate derived from the two body scattering theory for this potential, we have

$$D_c^+ f(q,p) = \int d^4 p_1 d^4 p_1' d^4 p' \dot{P}(p_1' p' \to p_1 p) f(q, p') f(q, p_1'),$$

$$D_c^- f(q,p) = \int d^4 p_1 d^4 p_1' d^4 p' \dot{P}(p_1 p \to p_1' p') f(q, p) f(q, p_1).$$

$$(8.162)$$

Furthermore these results can be put into terms of the experimentally measured scattering cross sections [Horwitz (1982)] in the form (we denote $q_1 - q_2$ by q_r, $\frac{1}{2}(p_1 - p_2)$ by p_r, and $P = p_1 + p_2$, and assume a narrow distribution over the mass shifts)

$$\partial_\tau f(q,p) + \frac{p^\mu}{M} \frac{\partial}{\partial q^\mu} f(q,p) = 4\pi \int d^3 p_r d^3 p_r' \frac{|\mathbf{p}_r'|}{M} \frac{d\sigma^{exp}}{d^3 p_r} (p_r' \to p_r; P)$$

$$\times \{ f(q,p') f(q, p_1') - f(q,p) f(q, p_1) \}.$$

$$(8.163)$$

With this final form of the Boltzmann equation, we can discuss the relativistic H-theorem. Defining the functional [Huang (1967)]

$$H(\tau) = \int d^4 q \, d^4 p \, f(q, p, \tau) \ln f(q, p, \tau) \equiv -S(\tau)/k_B, \qquad (8.164)$$

where $S(\tau)$ is the entropy. Then, the derivative of $H(\tau)$ is

$$\frac{dH}{d\tau} = \frac{1}{64} \int d^4 q \, d^4 p \, d^4 p_1 d^4 p_1' [\ln f(q, p_1) f(q, p) - \ln f(q, p_1') f(q, p')]$$

$$\times \{ f(q, p') f(q, p) f(q, p_1) \} \dot{P} \left(\left(\frac{p_1 - p}{2} \to \frac{p_1' - p'}{2} \right) ; P \right).$$

$$(8.165)$$

Since $\dot{P}(p_r \to p_r' : P) \geq 0$, and the remaining factor in the integrand is non-positive, we obtain

$$\frac{dH(\tau)}{d\tau} \leq 0, \qquad (8.166)$$

the relativistic H-theorem.

This result implies that the entropy $S(\tau)$ is monotonically increasing as a function of τ, but *not necessarily in t*, since the directions of t and τ for the antiparticle are opposite. In the nonrelativistic limit, the standard H theorem is recovered, since t and τ become identical.

8.9 Summary

In this chapter, the relativistic mechanics discussed in previous chapters has been applied to the construction of the covariant statistical mechanics of many body systems. It was assumed that a quasi-equilibrium state can be realized with a large number of events (particles) concentrated for a sufficient time in a bounded region of phase space, without introducing walls in space-time. With this assumption, one obtains the correct results for the statistical ensembles in the Galilean limit of the theory.

We discussed a high temperature phase transition leading to a restricted mass range. We then discussed second quatization of the fields and black body radiation. Finally, we discussed the covariant Boltzmann equation, for which, in the presence of antiparticles, admits a decrease in the entropy of an isolated system.

References*

Abraham, M., Ann. Phys. **10**, 105 (1903), *Theorie der Elektrizität*, vol.II, Springer, Leipzig, (1905); Lorentz, H.A., *The Theory of Electrons*, (Dover, New York (1952)[Lectures 1906; first edition appeared in 1909]; P.A.M Dirac, Proc. Roy. Soc. London **A 167**, 148 (1938).

Acquafredda, R., *et al.*, Jour. of Instr. **4**, P04018 (2009); Adam, T., *et al.*, arXiv 1109.4897 v4 (2011); Antonello, M., *et al.*, arXiv 1110.3763 (2011); arXiv 1203.3433 (2012).

Adam, T., *et al.* Jour. High Energy Phys. **1**, 153 (2013); arXiv 1212.1276 (2012).

Aharonov, Y. and Albert, D.Z., Phys. Rev. **D 24**, 359 (1982).

Aharonov, Y., personal communication, 1983.

Aharonovich, I. and Horwitz, L.P., Jour. Math. Phys. **51**, 052903 (2010); **52**, 082901 (2011); **53**, 032902 (2012). See also, I. Aharonovich and Horwitz, L.P., Jour. Math. Phys. **47**, 122902 (2006).

Amrein, W.O., Jauch, J.M., and Sinha, K.B., *Scattering Theory in Quantum Mechanics*, Benjamin, W.A., Reading, Mass. (1977).

Arnett, W.D., Bahcall, J.N., Kirschner, R.P., and Woolsey, S.E., Ann. Rev. Astron. Astrophys. **27**, 629 (1989). See also, Franson, J.D., arXiv:1111.6986; Longo, M.J., Phys. Rev. Lett. **60**, 173 (1988); Krauss, L.M. and Tremaine, S., Phys. Rev. Lett. **60**, 176 (1988).

Arshansky, R. and Horwitz, L.P., J. Phys. A: Math. Gen. **15**, 659 (1982).

Arshansky, R. and Horwitz, L.P., Phys. Rev. **D 29**, 2860 (1984).

Arshansky, R. and Horwitz, L.P., Found. Phys. **15**, 701 (1985).

Arshansky, R.I. and Horwitz, L.P., Jour. Math. Phys. **30**, 66, 380 (1989). See also, Arshansky, R.I. and L.P.Horwitz, Jour. Math. Phys. **30**, 213 (1989).

Ashtekar, A., Personal communication (1982).

---

*This list includes reference to related topics as well as those referred to in the text.

Ashtehkar, A., Personal communication (1982).

Bacry, M., personal communication (1990).

Bachar, Y., Arshansky, R., Horwitz, L.P., and Aharonovich, I., Found. of Phys. **44**, 1156 (2014).

Bahcall, J., *Neutrino Astrophysics*, Cambridge Univ. Press, Cambridge (1989).

Balescu, R., *Equilibrium and Nonequilibrium Statistical Mechanics*, Wiley, N.Y. (1975).

Bandyopadahyay, A., Choubey, S., Goswami, S., and Kamales, A., Phys. Rev. **D 65**, 073031 (2002).

Bargmann, V., Ann. Math. **48**, 568 (1947).

Bauer, M., arXiv 0908.2789 (2013).

Baumgartel, W., Math. Nachr. **75**, 173 (1976).

Baym, G., *Lectures on Quantum Mechanics*, Benjamin, W.A., N.Y. (1969).

Bekenstein, J.D., Phys. Rev. **D70**, 083509 (2004).

Bekenstein, J.D. and Milgrom, M., Astrophys. Jour. **286**, 7 (1984).

Bekenstein, J.D. and Sanders, R.H., Astrophys. Jour. **429**, 480 (1994).

Bennett, A., Jour. Phys. A: Math. Theor. **45**, 285302 (2012).

Bennett, A.F., Found. of Phys. **45**, 370 (2015).

Berry, M.V., Proc. Roy. Soc. (London), 45 **A 392** (1984).

Biedenharn, L.C. and Louck, J.D., *Angular Momentum in Quantum Physics* Encyclopedia of Math. and its Applications, vol. 8, Addison-Wesley, Reading (1981).

Dunos, M. and Biedenharn, L.C., Phys. Rev. **D 36**, 3069 (1987). See also, Yamagishu, N., Prog. Theor. Phys. **78**, 886 (1987).

Bjorken, J.D. and Drell, S.D., *Relativistic Quantum Mechanics*, McGraw Hill, New York (1964).

Bleistein, N., Neumann, H., Handelsman, R., and Horwitz, L.P., Nuovo Cimento **A 41**, 389 (1977).

Boerner, H., *Representations of Groups*, p. 312, North Holland, Amsterdam (1963).

Bogliubov, N.N. and Shirkov, D.V., *Introduction to the Theory of Quantized Fields*, Interscience, New York (1959).

Bohm, A. and Gadella, M., *Dirac Kets, Gamow Vectors, and Gel'fand Triplets, The Rigged Hilbert Space Formulation of Quantum Mechanics*, Springer, New York (1989).

Niels Bohr, Philosophical Magazine and Jour. of Science **26**, 151, London, Edinburgh and Dublin (1913).

Max Born, *Die Relativitäts-theorie Einsteins und ihre Physikalischen*, Grundlegen, Springer (1920).

Born, M., Einstein's Theory of Relativity, Dover, New York (1962).

Burakovsky, L., Horwitz, L.P., and Schieve, W.C., Phys. Rev. **D 54**, 4029 (1996).

Calderon, E., Horwitz, L.P., Kupferman, R., and Shnider, S., Chaos **23**, 013120 (2013).

Chew, G., *The Analytic S Matrix*, Benjamin, New York (1966).

Clebsch, A., *Theorie der binären algebraischen Formen*, Teubner, Leuipzig (1872); Gordan, P., *Über das Formensystem binärer Formen*, Teubner, Leipzig (1875). See also, Edmunds, A.R., *Angular Momomentum in Quantum Mechanics*, Princeton Univ. Press, Princeton (1957); Biedenharn, L.C. and Louck, J.D., *Angular Momentum in Quantum Physics, Theory and Application*, Encyclopedia of Mathematics and its Applications, Ed. Gian-Carlo Rota, vol. 8, Addison Wesley, Reading, Mass. (1981).

Cohen, E. and Horwitz, L.P., Hadronic Jour. **24**, 593 (2001).

Cohen, A.G. and Glashow, S.L., Phys. Rev. Lett. **107**, 181803 (2011).

Contaldi, C.R., Wiseman, T., and Withers, B., Phys. Rev. **D 78**, 044034 (2008).

Cook, J., Jour. Math. and Phys. **36**, 82 (1957).

Cook, J.L., Aust. Jour. Phys. **25**, 141 (1972).

Cooper, L., Phys. Rev. **104**, 1189 (1956).

Cornfield, I.P., Fomin, S.V., and Ya. Sinai, G., *Ergodic Theory*, Springer, Berlin (1982).

Currie, D.G., Jordan, T.F., and Sudarshan, E.C.G., Rev. Mod. Phys. **35**, 350 (1963).

Data Review of Particle Physics, Phys. Rev. **D 86**, 010001 (2014).

Davidson, M., Found. Phys **44**, 144 (2014).

Davisson, C. and Germer, L.H., Phys. Rev. **30**, 705 (1927).

Davisson, C. and Germer, L., Bell System Technical Journal **7**, 90–105 (1928).

de Broglie (1925). L. de Broglie, Ann. de Physique **3**, 22.

S.R. de Groot, W.A. van Leeuwen and Ch. G. van Weert, *Relativistic Kinetic Theory and Applications*, North Holland N.Y. (1980).

P. Di Francesco, Mathieu, P., and D. Sénéchal, *Conformal Field Theory*, Springer (1997).

Dirac, P.A.M., *Quantum Mechanics*, First edition, Oxford Univ. Press, London (1930), third edition (1947).

Dirac, P.A.M., Fock, V.A., and Podolsky, B., Phys. Zeits. f. Phys. **47**, 631 (1932). See also, Tomanaga, S., Nobel Lecture (1966). Dirac (1966). Dirac, P.A.M., *Lectures on Quantum Field Theory*, Belfer Graduate School, Yeshiva University, Academic Press, N.Y. (1966).

Dirac, P.A.M., Proc. Roy. Soc. (London) **A 136**, 453 (1932).

Dirac, P.A.M., *Lectures on Quantum Field Theory*, Graduate School of Science, Yeshiva University, New York, Academic Press N.Y. (1966).

Dixmeier, J., *Les C*-Algèbres et leurs Representations*, Gauthier-Villars, Paris (1969). See also, M. Dupré, Jour. Funct. Anal. **15**, 244 (1974).

Dothan, Y., M. Gell-Mann and Y. Ne'eman, Phys. Lett. **17**, 148 (1965). See also, Y. Ne'eman,*Algebraic Theory of Particle Physics*, W.A. Benjamin, New York (1967).

Eden, R.J., *High Energy Collisions of Elementary Particles* Cambridge Univ. Press, Cambridge (1967). See also, Eden, R.J., Landshoff, P.V., Olive, D.I., and Polkinghorne, J.C., *The Analytic S-Matrix*, Cambridge Univ. Press., Cambridge (1966).

Ehrenfest, P., Zeits. f. Physik **45**, 455 (1927).

Eimerl, D., Ann. Phys. (NY) **91**, 481 (1975). Horwitz(1981a). Horwitz, L.P. and Rotbart, F.C., Phys. Rev. **D 24**, 2127 (1981).

Einstein, A., Ann. Phys. (Germany) **17**, 891 (1905).

Einstein, A., Annalen der Physik **49**, 769 (1916).

Einstein, A., *The Meaning of Relativity*, Princeton University Press (Princeton) (1922).

Einstein, A., Podolsky, B., and Rosen, N., Phys. Rev. **48**, 696 (1935).

Eisenberg, E. and Horwitz, L.P., *Resonances, Instabilities and Irreversibility*, vol. XCIX, p. 245, Advances in Chemical Physics, ed. Prigogine, I. and Rice, S., John Wiley (1997); Horwitz, L.P. and Eisenberg, E., *Classical Quantum Correspondence*, Proc. 4th Drexel Symposium on Quantum Non-Integrability, Drexel University, Philadelphia, p. 267, ed. Feng, D.H. and Hu, B.L., International Press, N.Y. (1997); Horwitz, L.P., Eisenberg, E., and Strauss, Y., *Proc. of New Developments on Fundamental Problems in Quantum Physics*, Oviedo, Spain,*Fundamental Theories in Physics*, p. 171, **81**, Kluwer, Dordrecht. Strauss, Y., Eisenberg, E., and Horwitz, L.P., Jour. Math. Phys. **41**, 8050 (2000).

Horwitz, L.P. and Engelberg, E.Z., Phys. Lett. **A 374**, 40 (2009).

Faddeev, L.D. and Popov, V.N., Phys. Lett. **B 25**, 29 (1967).

Famaey, B. and McGaugh, S., Living Reviews in Relativity **10**, 15 (2012).

Fetter, A.L. and Walecka, J.D., *Quantum Theory of Many Particle Systems*, McGraw Hill, New York (1971).

Feynman, R.P., Phys. Rev. **76**, 769 (1949).

Feynman, R.P., Phys. Rev. **80**, 440 (1950).

Feynman, R.P., Kislinger, M., and Ravndal, F., Phys. Rev. **D 3**, 2706 (1971).

Flesia, C. and Piron, C., Helv. Phys. Acta **57**, 697 (1984).

Floquet, M.G., Ann. Ecole Norm. Suppl. **12**, 47 (1883); Cycon, H.L., Froese, R.G., Kirsch, W., and Simon, B., *Schrödinger Operators* *with Application to Quantum Mechanics and Global Symmetry*, p.146, Springer Varlag N.Y. (1987).

Fock, F.V., Phys. Zeits. Sowjetunion **12**, 404 (1937).

Fogli, G., Lisi, E., and Scioscia, G., arXiv hep-ph/0110307 (2002). See also, Kayser, B., SLAC-PUB-7123 (2005), Aaltonen, T., *et al.*, Phys. Rev. **D 85**, 012009 (2012), Lees, J.P., *et al.* Phys. Rev. Lett. **109**, 211801 (2012), Bhattacharya, B., Gronau, M., and Rosner, J.L., Phys. Rev. **D 85**, 054014 (2012).

Foldy, L.L. and Wouthuysen, S.A., Phys. Rev. **78**, 29 (1950).

Friedman, A., Zeits. fur Physik **A 21**, 326 (1924); Lemaitre, G., Monthly Notices of the Roy. Astro. Soc. **91**, 483 (1931)[trans. from Ann. de la Societe Scientifique de Bruxelles **A 47**, 49 (1927)]; see also Lemaitre, G., Ann. de la Societe Scientifique de Bruxelles **A 53**, 51 (1933); Robertson, H.P., Astrophys. Jour. **82**, 284 (1935), **83**, 187 (1936), **83**, 257 (1936).

Friedrichs, K.O., Comm. Pure Appl. Phys. Math. **1**, 361 (1950); Lee, T.D., Phys. Rev. **95**, 1329 (1956).

't Hooft, G. and Veltman, M., Nuc. Phys. **B44**, 189 (1972).

Galapon, E.A., Proc. Roy. Society A London, **458**, 451 (2002). See also, Aharonov, Y., Oppenheim, J., Popescu, S., Reznik, B., and Unruh, W.G., Lecture Notes in Physics **517**, 204 (1999).

Galileo Galilei, *Dialog Concerning the Two Chief World Systems* (1632).

Gamow, G., Zeits. f. Phys. **51**, 204 (1928).

Gel'fand, I.M., Minlos, R.A., and Ya, Z., Shapiro, *Representations of the Rotation and Lorentz Groups and Their Applications*,Pergamon Press, New York (1963).

Gell-Mann, M., Phys. Rev. **125**, 1067 (1962).

Gershon, A. and Horwitz, L.P., Jour. Math. Phys. **50**, 10274 (2009); Horwitz, L.P., Gershon, A., and Schiffer, M., Found. of Phys. **41**, 141 (2010).

Goldberger, M.L. and Watson, K.M., *Collision Theory* Wiley, N.Y. (1964).

Goldstein, H.H., *Classical Mechanics*, Addison, Wesley, N.Y. (1951).

Gordon, W., *Zeits. Fur Physik* **40**, 117 (1926).

Gottfried, K., *Quantum Mechanics*, Benjamin, New York (1966).

Gottfried, K. and Weisskopf, V., *Concepts of Particle Physics*, Oxford Univ. Press, Clarendon, Oxford (1986).

Green, M.B., Schwartz, J.H., and Witten, E., *Superstring Theory*, Cambridge Univ. Press, Cambridge (1986).

Haber, H.E. and Weldon, H.A., Phys. Rev. Lett. **46**, 1497 (1981); Phys. Rev. **D 25**, 502 (1982).

Hahne, G.E., J. Phys. A **367**, 144 (2003).

Hakim, R., *Introduction to Relativistic Statistical Mechanics*, World Scientific, Singapore (2011) see also, Hakim, R., and Mangeney, A., Lett. Nuovo Cim. **1**, 429 (1969).

Havas, P., Bull. Am. Phys. Soc. **1**, 337 (1956).

Heitler, W., *Quantum Theory of Radiation*, Oxford (1936).

Henley, E.M. and Garcia, A., *Subatomic Physics*, World Scientific, Singapore (2007).

Henneaux, M. and Teitelboim, C., *Quantization of Gauge Systems*, Princeton Univ. Press, Princeton, N.J. (1992).

Hofstadter, R., Bumiller, R., and Yearian, M.R., Rev. Mod. Phys. **30**, 482 (1958).

't Hooft, G., Nuc. Phys. **B33**, 173 (1971).

Horwitz, L.P. and Marchand, J.P., Rocky Mountain Jour. of Math. **1**, 225 (1971).

Horwitz, L.P. and Piron, C., Helv. Phys. Acta **46**, 316 (1973).

Horwitz, L.P. and Piron, C., Helv. Phys. Acta **66**, 316 (1973). See also Collins, R.E., and Fanchi, J.R., Nuovo Cim. **48A**, 314 (1978); Fanchi, J.R., *Parametrized Relativistic Quantum Theory*, Kluwer, Dordrecht (1993).

Horwitz, L.P. and Rabin, Y., Lettere al Nuovo Cimento **17**, 501 (1976).

Horwitz, L.P. and Sigal, I.M., Helv. Phys. Acta **51**, 685 (1978).

Horwitz, L.P. and Soffer, A., Helv. Phys. Acta **53**, 112 (1980). See also, Soffer, A., Lett. Math. Phys. **8**, 517 (1984).

Horwitz, L.P., Schieve, W.C., and Piron, C., Ann. of Phys. **137**, 306 (1981). See also, Horwitz, L.P., Shashoua, S., and Schieve, W.C., Physica **A 161**, 300 (1989).

Horwitz, L.P., Schieve, W.C., and Piron, C., Ann. Phys. **137**, 306 (1981); Horwitz, L.P., Shashoua, S., and Schieve, W.C., Physica **A 161**, 300 (1989). See also, Burakovsky, L. and Horwitz, L.P., Found. Phys. **25**, 1335 (1995).

Horwitz, L.P. and Lavie, Y., Phys. Rev. **D 26**, 819 (1982).

Horwitz, L.P. and Rohrlich, F., , Phys. Rev. **D 26**, 3452 (1982).

Horwitz, L.P., Arshansky, R.I., and Elitzur, A., Found. Phys. **18**, 1159 (1988).

Horwitz, L.P., Found. Phys. **25**, 39 (1995).

Horwitz, L., Y. Ben Zion, Lewkowicz, M., Schiffer, M., and Levitan, J., Phys. Rev. Lett. **98**, 234301 (2007).

Horwitz, L.P., Gershon, A., and Schiffer, M., Found. Phys. **41**, 141 (2010). See also, Gershon, A. and Horwitz, L.P., Jour. Math. Phys. **50**, 102704 (2009).

Horwitz, L., Jour. Pys. A: Math and Theor. **46**, 035305 (2013).

Horwitz, L.P. and M. Zeilig-Hess, to be published (2015).

Horwitz, L.P. and A.Yahalom, to be published (2015).

Horwitz, L.P., Piron, C., and Reuse, F., Helv. Phys. Acta **48**, 546 (1975).

Howland, J.S., Math. Ann. **207**, 315 (1974); see also, Howland, J.S., Indiana Math. Jour. **28**, 471 (1979).

Hughston, L.P., Proc. Roy. Soc. London A **452**, 953 (1996). Brody, D.C. and Hughston, L.P., Jour. Phys. A: Math. Gen. **21**, 2885 (2006).

Hu, B.L., Lin, S.Y., and Louka, J., Class. and Quant. Grav. **29**, 224005 (2012).

Huang, K., *Statistical Mechanics*, Wiley, N.Y. (1967).

Hughston, L.P., Proc. Roy. Soc. of London A **452**, 953 (1996); Brody, D.C., ansd Hughston, L.P., Proc. Roy. Soc. of London **454**, 2445 (1998); Brody, D.C. and Hughston, L.P., Jour. of Geom. and Phys. **38**, 19 (2001).

Israel, W. and Kandrup, H.E., Ann. Phys. **182**, 30 (1984); Kandrup, H.E., Astr. and Space Sci. **124**, 359 (1986).

Itzykson, C. and Jean-Bernard Zuber, *Quantum Field Theory*, McGraw-Hill, New York (1980).

Jabs, A., Found. of Phys. **40**, 776 (2010).

Jackiw, R., Int. J. Mod. Phys. A **03**, 285 (1988).

Jackson, J.D., *Classical Electrodynamics*, 2nd edition, John Wiley H, New York (1974).

Jacob, M. and Wick, G.C., Ann. Phys. (NY) **7**, 404 (1959).

Jauch, J. and Rohrlich, F., *The Theory of Photons and Electrons*, Addison-Wesley, Cambridge, Mass. (1955). Second edition, Springer, New York (1976).

Jauch, J.M., *Foundations of Quantum Mechanics*, Addison-Wesley, Reading (1968).

Jones, H.F., *Groups, Representations and Physics*, Institute of Publishing, Bristol and Philadelphia (1990).

Jordan, T., personal communication (1980).

Jüttner, F., Ann. Phys. (Leipzig) **34**, 856 (1911).

Kaku, M., *Quantum Field Theory*, Oxford Univ. Press, New York (1993).

Kaluza, T., Sitz. Preuss. Akad. Wiss. Berlin, Math. Phys., 966 (1921); Klein, O., Z. f. Phys. A **37**, 895 (1926).

Kato, T., *Perturbation Theory for Linear Operators*, Springer, Berlin (1980).

Kayser, Phys. Lett. B *592*, 145 (2004); Phys. Rev. D **24**, 110 (1981).

Kim, Y.S. and Noz, M.E., Prog. Theor. Phys. **57**, 1373 (1977).

Kirsten, K., Class. Quantum Grav. **8**, 2239 (1991); J. Phys. A **24**, 3281 (1991). See also, Toms, D.J., Phys. Rev. Lett. **69**, 1152 (1992); Phys. Rev. D **47**, 2453 (1993).

Klein, O., *Zeits. fur Physik* **37**, 895 (1926).

Kostant, B., *Quantization and Unitary Representations* in Modern Analysis and Applications, Lecture Notes in Mathematics **170**, 87, Springer, Heidelberg (1970).

Kostant, B. and Sternberg, S., Ann. Phys. **176**, 49 (1987). See also, D. Hüsemoller, *Basic Bundle Theory and K-Cohomolgy Invariants*, Springer, Berlin (2008).

Kulander, K.C. and Lewenstein, M., *Atomic, Molecular and Optical Physics Handbook* (G.W. Drake, ed.), p. 828, American Institute of Physics Press, Woodbury, N.Y. (1996).

Lamb, W.E. and Retherford, R.C., Phys. Rev. **72**, 241 (1947).

Land, M., Shnerb, N., and Horwitz, L.P., Jour. Math. Phys. **36**, 3263 (1995).

Land, M., Found. of Phys. **28**, 1479 (1998).

Land, M., Jour. of Phys. Conf, Series **330**, 012014 (2011).

Landau, L. and Peierls, R., Z. Phys. **69**, 56 (1931).

Landau, L. and Lifshitz, E., *Classical Theory of Fields*, trans. M. Hamermesh, Addison Wesley, Cambridge, Mass. (1951).

Landau, L.D. and Lifshitz, *Quantum Mechanics*, Pergamon Press, Oxford (1965).

Lavie, Y., Personal communication (1982).

Lax, P.D. and Phillips, R.S., *Scattering Theory*, Academic Press, New York (1967). See also, Ja. Sinai, Izv. Akad. Nauk SSSR *Ser. Mat.* **25**, 899 (1961).

Lebedev, Pav., Ann. der Physik, Vierte Folge B 29.

Lee, T.D. and Yang, C.N., Phys. Rev. **104**, 254 (1956). See also, Lee, T.D., Oehme, R., and Yang, C.N., Phys. Rev. **106**, 340 (1957); Lee, T.D. and Wu, C.S., Ann. Rev. Nuc. Science (1966),511.

Lehmann, H., Symanzik, K., and Zimmermann, W., Nuovo Cimento **1**, 1425 (1955).

Leutwyler, H. and Stern, J., Phys. Lett. **B 69**, 207 (1977).

Levinson, N., Kgl. Danske Videnskab Mat.-fys Medd. **25**, No. 9 (1949).

Liko, T., Phys. Lett. **B 617**, 193 (2005).

Lin, S.Y. and Hu, B.L., Phys. Rev **D 79**, 085020 (2009).

Lindner, F., Schätzel, M.G., Walther, H., Baltuska, A., Goulielmakis, E., Krausz, F., Milošević, D.B., Bauer, D., Becker, W., and Paulus, G.G., Phys. Rev. Lett. **95**, 040401 (2005). See also, Horwitz, L., Phys. Lett. **A 355**, 1 (2006).

Liozier, J.T. and Maloney, J.R., Entropy **15**, 177 (2013).

Lippmann, B.A. and Schwinger, J., Phys. Rev. Lett. **79**, 469 (1950).

Llosa, J., (editor), *Relativistic Action at a Distance: Classical and Quantum Aspects*, Lecture Notes in Physics **162**, Springer, Berin (1982). See also, Todorov, I., SISSA Report, Trieste (1980) (unpublished), Horwitz, L.P. and Rohrlich, F., Phys. Rev. **D 24**, 1528 (1981); Iranzo, V., Llosa, J., Molina, A., and Marques, F., Ann. Phys. (NY) **150**, 1124 (1983); Todorov, I., Institute of Nuclear Research, Dubna Report E2-10175 (unpublished)(1976); Komar, A., Phys. Rev. **D 18**, 1881, 1887, 3617

(1978); Ph. Droz-Vincent, Phys. Ser. **2**, 129 (1970), Nuovo Cim. **58A**, 355 (1980); Crater, H. and Van Alstine, P., Ann. Phys. (NY) **148**, 57 (1983); Horwitz, L.P. and Rohrlich, F., Phys. Rev. **D 31**, 932 (1985).

Lorentz, H.A., Zitterungverslagen der Akad van Wettenschappen I, 74 (1892).

Ludwig, G., *Foundations of Quantum Mechanics I*, Springer-Verlag, New York (1982); *Foundations of Quantum Mechanics II*, Springer-Verlag, New York (1983).

Mackey, G.M., *Induced Representations of Groups and Quantum Mechanics*, W.A. Benjamin, New York (1968).

Mahan, G.D., *Many Particle Physics*, Plenum Press, New York (1990), 2nd printing (1993).

Maslov, V.P., *Operational Methods*, tr. Golo, V., Kulman, N., and Voropaeva, G., Mis, Moscow (1976). See also, Arnold, V., *Mathematical Methods of Classical Mechanics*, Springer-Verlag, New York (1974).

Maxwell, James Clerk, Philosophical Transactions of the Royal Society of London **115**, 459–512 (1865).

Merzbacher, E., *Quantum Mechanics*, 2nd edition, Wiley, New York (1970).

Michel, L., discussions at IHES, Bures-sur Yvette (1961).

Michelson, A.A. and Morely, E.M., American Jour. of Science **34**, 203 (1888).

Milgrom, M., Asrophys. Jour. **270**, 365, 371, 384 (1983). See Bekenstein, J.D., Contemporary Physics **47**, 387 (2006) for a review, references, and further development.

Miller, F.R. and Curtis, W.D., *Differential Manifolds and Theoretical Physics*, Aca. Press N.Y. (1986).

Misner, C.W., Thorne, K.S., Wheeler, J.A., *Gravitation*, Freeman, W.H., New York (1970).

Misra, B., Sudarshan, E.C.G., Jour. Math. Phys. **18**, 765 (1977); see also, Chiu, C.B., B,. Misra and Sudarshan, E.C.G., Phys. Rev. **D 16**, 520 (1977).

Moshinsky, M., Phys. Rev. **88**, 625 (1952).

Ed. G. Muga, Ruschhaupt, A., and Campo, A., *Time in Quantum Mechanics*, vol. 2, Springer Lecture Notes in Physics 789, Springer, Heidelberg (2009). See also, Ed. Muga, G., R. Sala Mayato and Egusquiza, I., *Time in Quantum Mechanics*, vol. 1, Springer Lecture Notes in Physics 734, Springer, Heidelberg (2008).

Naimark, M.A., *Linear Representations of the Lorentz Group* Pergamon, New York (1964).

Nambu, Y., Prog. Theor. Phys. **5**, 82 (1950).

Nambu, Y., in *Symmetries and Quark Models*, p. 269, ed. R. Chaud, Gordon and Breach, N.Y. (1970).

Ne'eman, Nuc. Phys. **26**, 222 (1961).

Isaac Newton, *Philosophia Naturalis Principia Mathematica*, London (1687); Cohen, I.B. and Whitman, A., *The Principia: Mathematical Principles of Natural Philosophy: A New Translation*, University of California Press, Berkeley (1999).

Newton, T.D. and Wigner, E., Rev. Mod. Phys. **21**, 400 (1949).

Newton, R.G., *Scattering Theory of Waves and Particles*, McGraw Hill, New York (1967).

Nichol, E., Vried Ann. **69**, 406 (1897).

Nunokawa, H., CBPF School (2006).

Oron, O. and Horwitz, L.P., Found. of Phys. **31**, 951 (2001).

Overduin, J.M. and Wesson, P.S., *The Light Dark Universe; Light from Galaxies, Dark Matter and Dark Energy*, World Scientific, Singapore (2008).

Palacios, A., Rescigno, T.N., and McCurdy, C.W., Phys. Rev. Lett. **103**, 253001 (2009).

Pauli, W., *Relativitätstheorie*, Encyk. der Mat. Wiss. **V19**, B.C. Teubner, Leipzig (1921); trans:*Theory of Relativity*, Pergamon Press (1958).

Pauli, W. and Weisskopf, V., Helv. Phys. Acta **7**, 709 (1934).

Gerhard Paulus, Personal communication (2005).

Peskin M.E. and Schroeder, D.V. *An Introduction to Quantum Field Theory*, Addision-Wesley, Reading, Mass. (1995).

Piron, C., *Foundations of Quantum Physics*, Benjamin, W.A., Reading (1976).

Polchinski, J., *String Theory I,II*, Cambridge Univ. Press, Cambridge (1998).

Poynting, J.H., Phil. Trans. of the Royal Society of London **175**, 343 (1881).

Racah, G., Phys. Rev. **62**, 438 (1942); **63**, 367 (1943).

Raimes, S., *The Wave Mechanics of Electrons in Metals*, North Holland, Amsterdam (1961).

Rarita, W. and Schwinger, J., Phys. Rev. **60**, 61 (1941).

Reed, M. and Simon, B., *Methods of Modern Mathematical Physics, III, Scattering Theory*, Academic Press, New York (1979).

Rohrlich, F., Phys. Rev. **D 23**, 1305 (1981).

Saad, D., Horwitz, L.P., and Arshansky, R.I., Found. of Phys. **19**, 1125 (1989).

Salières, P., Carré, B., Le D'eroff, L., Grasbon, F., Paulus, G.G., Walther, H., Kopold, R., Becker, W., Milosevic, D.B., Sanpera, A., and Lewenstein, M., Science **292**, 902 (2001).

Schieve, W.C. and Horwitz, L.P., *Quantum Statistical Mechanics*, Cambridge University Press, Cambridge (2009).

Schrödinger, E., Ann. Physik, **81**, 109 (1926); Gordon, W., Physik, Z., **40**, 117 (1926); Klein, O., Physik, Z., **41**, 407 (1927).

Schwarzschild, K., Sitz. der Deutschen Akademie der Wissenschaften zu Berlin, Klasse for Mathematik, Physik, und Technik, 189 (1916).

Schweber, S.S., *An Introduction to Relativistic Quantum Field Theory*, Harper and Row, New York (1964).

Schwinger, J., Physical Review **82**, 664 (1951).

Schwinger, J., *Quantum Dynamics and Kinematics*, Frontiers in Physics. N.Y. (2000).

Shapere, A. and Wilczek, F., *Geometric Phases in Physics*, Advanced Series in Mathematical Physics, vol. 5, World Scientific, Singapore (1989).

Shapere, A. and Wilczek, F., *Geometric Phases in Physics*, World Scientific, Singapore (1989).

Shikerman, F., A. Pe'er and Horwitz, L.P., Phys. Rev. **A 84**, 012122 (2011).

Shnerb, N. and Horwitz, L.P., Phys. Rev. **A 48**, 4068 (1993).

Sigal, I.M. and Soffer, A., Ann. Math. **126**, 35 (1987).

Silman, J., Machness, S., Shnider, S., Horwitz, L.P., and Belenkiy, A., Jour. of Phys.: Math. and Theory **41**, 255303 (2008).

Sklarz, S. and Horwitz, L.P., Found. Phys. **31**, 909 (2001).

Sommerfeld, A., *Atombau und Spektrallinien*, Vieweig and Sohn Braunschweig, (1921). See also, Ghoshal, S.K., Nandi, K.K., and Ghoshal, T.K., Indian J. Pure and Appl. Math., **18**, 194 (1987); Bergmann, P.D., *Introduction to the Theory of Relativity*, Prentice-Hasll, Engelwood Cliffs, N.J.

Sommerfeld, A., *Atombau und Spektrallinien*, vol. II, Chap. 4, Friedrich Vieweg und Sohn, Braunschweig (1939).

Strauss, Y. and Horwitz, L.P., Found. of Phys. **30**, 653 (2000).

Strauss, Y. and Horwitz, L.P., Found. of Phys. **30**, 653 (2000). See also Horwitz, L.P. and Strauss, Y., Found. of Phys. **28**, 1607 (1998).

Strauss, Y., Horwitz, L.P., and Eisenberg, E., Jour. Math. Phys. **41**, 8050 (2000).

Strauss, Y. and Horwitz, L.P., Jour. Math. Phys. **43**, 2394 (2002).

Strauss, Y., Silman, J., Machnes, S., and Horwitz, L.P., Comptes Rendus Mathematique **349**, 1117 (2011).

Stueckelberg, E.C.G., Helv. Phys. Acta **14**, 372, 585; **15**, 23 (1942).

Stueckelberg, E.C.G., Helv. Phys. Acta **14**, 372, 588 (1941); **15**, 23 (1942).

Sudarshan, E.C.G. and Mukunda, N., *Classical Dynamics. a Modern Perspective*, Wiley, New York (1974)

Sudarshan, E.C.G., private communication (1981).

Sudarshan, E.C.G., Mukunda, N., and Goldberg, J.N., Phys. Rev. **D 23**, 2218 (1981).

Suleymanov, M., Horwitz, L., and Yahalom, A., *Covariant spacetime string*, submitted (2015).

Synge, J.L., *The Relativistic Gas*, North Holland, Amsterdam (1957).

Taylor, J.R., *Scattering Theory: The Quantum Theory on Nonrelativistic Collisions*, Wiley, New York (1972).

Tomczak, S.P. and Haller, K., Nuovo Cimento **B 8**, 1 (1972); Haller, K. and Sohn, R.B., Phys. Rev **A 20**, 1541 (1979); Haller, K., Acta Physica Austriaca **42**, 163 (1975); Haller, K., Phys. Rev. **D 36**, 1830 (1987).

Tomonaga, S., Phys. Rev. **74**, 224 (1948); Schwinger, J., Phys. Rev. **73**, 416 (1948); Phys. Rev. **74**, 1439 (1948).

Trump, M.A. and Schieve, W.S., *Classical Relativistic Many-Body Dynamics*, Kluwer, Dordrecht (1999).

Tulley, R.B. and Fisher, J.R., Astron. Astrophys. **54**, 661 (1977).

Turyshev, S.G., Toth, V., Kellogg, L., Lau, E., and Lee, K., Int. Jour. Mod. Phys. **D 15**, 1,(2006). Turyshev, S.G., Toth, V., Kinwella, G., S.-C. Lee, Lok, S., and Ellis, J., Phys. Rev. Lett. **108**, 241101 (2012).

Van Hove, L., Proc. Roy. Aca. Belgium **26**, 1 (1951). See also Groenwold, H.J., Physica **12** 405 (1946). Konstant (1970).

von Neumann, J., *Mathematical Foundations of Quantum Mechanics*, Princeton Univ. Press, Princeton (1955).

von Neumann, J., *Mathematical Foundations of Quantum Mechanics*, Princeton Univ. Press, Princeton (1971).

Weinberg, S., Phys. Rev. Lett. **19**, 1264 (1967); A. Salam *Elementary Particle Theory: Relativistic Groups and Analyticity*, 8th Nobel Symposium, ed. Svartholm, N., Almquist and Wiksell, 367, Stockholm (1968); Glashow, S., Iliopoulis, J., and Maini, L., Phys. Rev. **D 2**, 1285 (1970).

Weinberg, S., *Gravitation and Cosmology: Principles and Applications of the General Theory of Relativity*, John Wiley and Sons, New York (1972).

Weinberg, S., *Quantum Theory of Fields, Volume 1, Foundations* Cambridge Univ. Press, Cambridge (2003). See also Zee, A., *Quantum Field Theory in a Nutshell*, Princeton (2003).

Weisskopf, V.F. and Wigner, E.P., Zeits. f.Phys. **63**, 54 (1930).

Wesson, P.S., *Five Dimensional Physics, Classical and Quantum Consequences of Kaluza-Klein Cosmology*, World Scientific, Singapore (2006).

Weyl, H., *Gruppentheorie und Quantenmechanik*, 2nd ed. Hirzel, Leipzig (1931); trans. *The Theory of Groups and Quantum Mechanics*, Dover, N.Y.(1950).

Weyl, H., *Space, Time, Matter*, Dover, New York (1952).

Wick, G.C., Wightman, A.S., and Wigner, E.P., Phys. Rev. **88**, 101 (1952).

Wigner, E.P., *Gruppentheorie und ihre Anwendung auf die Quantenmechanik der Atomspektren*, Fried Viewag & Sohn, Akt. Ges.,

Braunschweig (1931); trans. *Group Theory and its Application to the Quantum Mechanics of Atomic Spectra*, Aca. Press, N.Y. (1959).

Wigner, E., Ann. of Math. **40**, 149 (1939). See also, G.W. Mackey, *Induced Representations of Groups and Quantum Mechanics*, Benjamin, New York (1968); Bargmann, V., Ann. of Math. **48**, 568 (1947).

Wigner, E.P., Phys. Rev. **98**, 145 (1955).

Wigner, E.P., *Gruppentheorie und ihre Anwendung auf die Quantenmechanik der Atomspektra*, Friedr. Vieweg & Sohn, Alet. Ges., Braunshweig (1931).

Wigner, E.P. (1931). See also, Weinberg, S. (1995). Weinberg, S., *Quantum Field Theory I*, p. 49, Cambridge Univ. Press, Cambridge (1995).

Winternitz, P., Personal communication (1985).

Wu, T.T. and Yang, C.N., Phys. Rev. **D 12**, 3845 (1975). See also, Wu, A.C.T. and Yang, C.N., Int. Jour. Mod. Phys. **A 21**, 3235 (2006).

Yang, C.N. and Mills, R., Phys. Rev. **96**, 191 (1954).

Zaslavsky, G.M., *Chaos in Dynamic Systems*, Harwood Aca. Pub. New York (1985). See also, *Physics of Chaos in Hamiltonian Dynamics*, Imperial College Press, London (1998), *Hamiltonian Chaos and Fractal Dynamics*, Oxford Univ. Press, Oxford (2005).

Zmuidzinas, J.S., Jour. Math. Phys. **7**, 764 (1966).

Printed in the United States
by Baker & Taylor Publisher Services